WATER AND ENVIRONMENTAL MANAGEMENT SERIES

Decentralised Water and Wastewater Systems

International Conference
Fremantle, Western Australia
10 - 12 July 2006

Edited by
Dr Kuruvilla Mathew
Dr Stewart Dallas
Prof Goen Ho

With the support of

UNESCO IWA Publishing Murdoch UNIVERSITY

Published by IWA Publishing, Alliance House,
12 Caxton Street, London SW1H 0QS, UK

Telephone: +44 (0) 20 7654 5500
Fax: +44 (0) 20 7654 5555;
Email: publications@iwap.co.uk
Web: www.iwapublishing.com

ISBN: 978-1-84339-515-7

Preface

The International Conference on Decentralised Water and Wastewater Systems 2006 was organised by the Environmental Technology Centre at Murdoch University. It aimed to promote the concept of decentralised systems and to encourage partnerships with industry, consultants and academics.

The World Health Organization estimates that globally one billion people are without access to safe water and two billion people without adequate sanitation. The health consequence of such a situation is very serious and demands urgent appropriate action.

The centralised system, which is still promoted all over the world, will not help in providing water and sanitation for all, as it needs huge financial and technological resources. The Conference promoted decentralised systems as a means of solving this problem.

There were two keynote speakers, 58 oral presentations and 20 poster presentations at the Conference.

All the papers submitted at the Conference have undergone through a review process. The selected papers are presented in this publication.

The authors wish to acknowledge the financial assistance of the Australian National Commission of UNESCO for the publication of the proceedings.

Dr Kuruvilla Mathew
Dr Stewart Dallas
Prof Goen Ho
Editors

Environmental Technology Centre, Murdoch University, Perth, Western Australia
The Centre is a Centre of Excellence in Industry Focussed Research & Development

Table of Contents

1 Irrigation and Plant Growth

1.1 Solar-powered saline groundwater management for sustainable agriculture

Ahmed Hasson[1] and Dallas, S.
[2]UNEP-IETC Environmental Technology Centre, Murdoch University,
Perth 6150 Western Australia
a.hasson@murdoch.edu.au

Abstract

Many rural areas in Iraq suffer from saline groundwater, which is limiting agricultural development. The latitude of this study site has a similar climate to the Moora to Katanning region of Western Australia in the south west wheat belt. Other similarities include rising saline groundwater levels, low winter rainfall, and high evaporation with a high sun-shine duration. A strategy incorporating solar pumping of saline groundwater to shallow ponds incorporating a novel evaporation reduction technique, and subsequent dilution with winter rainfall to produce brackish quality water suitable for livestock and irrigation has been developed in Alexandria and is described in this paper.

The site for this research is situated at 32.00 N; 44.35 E in rural of Alexandria, where the national electricity grid is not available to power groundwater pumps. Solar pumping from a depth of 14m is used to fill three interconnected ponds of approximately 115m diameter and 0.35 to 2.6m in depth. Approximately 130,000m^3 of saline groundwater (1,200mS/m, 6600mg/L) is pumped to the ponds each season which when diluted with average winter rainfall of 500-650mm/year provides a resultant shandy of water which can be classified as moderate/brackish (850 mS/m). This is suitable for irrigation of date palms and alfalfa crops and for livestock. The system is made viable through the innovative use of date palm leaves which are floated out over the ponds as a mat and reduce evaporation by over 80% of net pan evaporation.

Key Words

Evaporation, Saline groundwater, Solar pumping, Western Australia, Iraq

Introduction

Groundwater is a contributing factor to the transport of salinity and therefore has the potential to impact irrigated agriculture. In addition water leakage below irrigated farming systems is a significant contributing factor to the rise in the groundwater table and it may simultaneously transport nutrients, particularly the nitrogenous forms.

In the southern region of Iraq water scarcity is commonplace, especially during the dry season periods. Evaporation plays an important role in the region as it increases the saturated salts in ponds and causes salts to accumulate in the soil, especially in hot, windy conditions (Hasson and AlSajir, 1989). The soil is further impacted as a result of its inherent properties as well as geological drainage patterns which have resulted in the storage of large amount of salts in the lands.

Salinity control can however be achieved via manipulation of the groundwater table and this is done successfully by instigating water pumping. Water for irrigation by groundwater pumping in these areas is required for agriculture however no national electric grid or water network is available. Consequently the most suitable approach is solar water pumping using photovoltaic (PV) powered pumps, (Hasson and AlSajir, 1989). On-site water pumping with storage is the most inexpensive method however the viability of this option is dependent upon the local topography. Singh and Christian (1999) studied basin storage options with various sizes ranging from 2-30ha, considered as a small basin, and 30-200ha considered as a large basin.

Leany and Christian (2000) reported however that leakage from such unlined reservoirs can vary from 0.5–5.4mm/day with acceptable losses set as 0.5 – 1.0mm/day. Leakage of more than 3mm/day is considered unviable.

Methods

Site Description

Iraq is characterised by an arid to semi-arid climate being driest in the Eastern region and ranging to mild in the Mesopotamian parts. Rainy seasons are short and the annual mean rainfall varies with geographical region from 500 to 650 mm/year. The region described in this paper is characterised by clear skies throughout most of the year with potential evaporation rates up to 17.5 to 25.2 mm/day in summer and monthly solar radiation as shown in Figure 1.1.1. Daily micro-climatological parameters were measured at Alexandria (32.00 N; 44.35 E). These are regions where the national electricity grid is not available to power groundwater pumps. The potential for development of a moderately saline water final product from saline groundwater pumped from depths of up to 14m using solar energy was studied.

Water Pumping

The water table was tested using piezometers (5 cm diameter) which were drilled to full depth.

The photovoltaic surface area was $27m^2$ and calculated as follows:

$$S= P/(R_s\eta_c) \tag{1.0}$$

Where S is the surface area of the photovoltaic system (m), R_s is total solar radiation, ($MJ/m^2/d$), η_c is the photovoltaic efficiency (10%) and P is the pump power.

The pumping (hydraulic) energy required to deliver a volume of water is given by:

$$E_h= \rho_w gVH \tag{2.0}$$

Where Eh is the required hydraulic energy, V the required volume of water, H the total head, Úw the density of water and g the gravitational acceleration.

The input energy for the pumping system undergoes several conversions before it is made available as useful hydraulic energy. The input energy requirement for pumping is generally much greater than the useful hydraulic energy output because of the energy losses associated with each conversion. The power required to lift a given quantity of water depends on the length of time that pump is used and represented in Figure 1.1.2. It is given in the form:

$$P_h= \rho_w gHQ \tag{3.0}$$

The total head comprises the sum of the static head and the head loss in the pipes, which depends on the pipe diameter and flow rate (Q).

An automatic data acquisition control (Model 3054 A) was synchronized to provide the components of the micro-climate: evaporation rate, heat flux, thermal conductivity, specific heat net radiation, total solar radiation, ambient air temperature, rainfall, humidity, at hourly intervals.

Ponds and Topography

Three "artificial" connecting ponds of identical cross-section, each pond size approximately 95, 114 and 120m in diameter are located at the farmland of interest where the land slope is approximately 4.0–4.5 per cent. The deepest point of the ponds is 2.8m ranging to 0.35m at the shallowest point.

The ponds are considered to be surface catchments as they collect the surrounding area rainfall water. The inflow from pumped water and rainfall (surface run-off water) is less than the ponds' volume. The ponds' soil type is a silty clay of a low permeability. The bases of the ponds were compacted with a heavy compacting machinery to prevent water leakage and the ultimate leakage rate was estimated as less than 1 per cent.

The average conductivity (EC_e) in the "A" horizon is >1000mS/m (>5500mg/L) and indicates a high salt content in the experimental site's soils. During the research period the groundwater table was monitored for both depth and salinity using piezometers.

The effective soil water content as well as the soil bulk density were determined gravimetrically from undisturbed 7.6cm diameter core samples taken from the ponds. The specific heat of the soil was determined as 0.35 cal/gm/day.

The ponds receive approximately 130,000m^3 from pumped water of approximately 1,200mS/m which is classified saline. Added to this is the rainfall contribution (surface run-off) water of 450 – 650mm/year which dilutes this saline water to achieve a final water quality of approximately 850mS/m which is classified as moderately brackish water. This final quality water is suitable for brackish water tolerant crops such as date-palm, alfalfa, clover, barley, sorghum, and is suitable for stock.

The ponds are covered with floating date-palm leaves which are woven into a light textile mat (2.30 -3.50mm thick) to reduce the potential evaporation rates (Figure 1.1.4). Evaporation from the ponds was measured by recording water depth using sensitive transducers placed at a height of 30cm above the pond and measured at weekly intervals. The electrical conductivities of the water were measured on different sides of the ponds and at various depths also at weekly intervals.

Economic Assessment

Solar PV pumps and reuse of pumped water for agricultural purposes are technically viable but where alternatives exist the evaluation of the alternatives must include both economic and technical considerations. There are three common techniques that are used for making an economical evaluation: payback period, rate of return and life cycle costs. The purpose of the cost appraisal is to calculate the unit water cost. The methodology that is used to compare the costs of the major alternative pumping and evaporative ponds methods involves the use of a discount rate which reflects the cost of the capital (Hammad, 1999).

The Analytic Hierarchy Process (AHP) can be used as a practical approach for addressing these cost-benefit application issues (Saaty 1980). It has been effective in structuring many types of complex multi-criterion problems and the AHP has been applied to the evaluation of energy systems, selection of R&D programs and water polices (Gocht and Gillett, 1985; Gocht, 1993). In a forthcoming paper the Analytic Hierarchy Process (AHP) will be used for studying the cost-benefit applications of solar water pumping systems and reuse of discharged saline water and mixing with rainwater for agricultural applications.

Results and Discussion

The monthly average daily total solar radiation and net solar radiation recorded for the above site is presented in Figure 1.1.1. This illustrates the potential for this site in terms of solar irradiance. The annual average total solar radiation and net solar radiation densities are 23.3 and 15.8MJm^2 /day respectively.

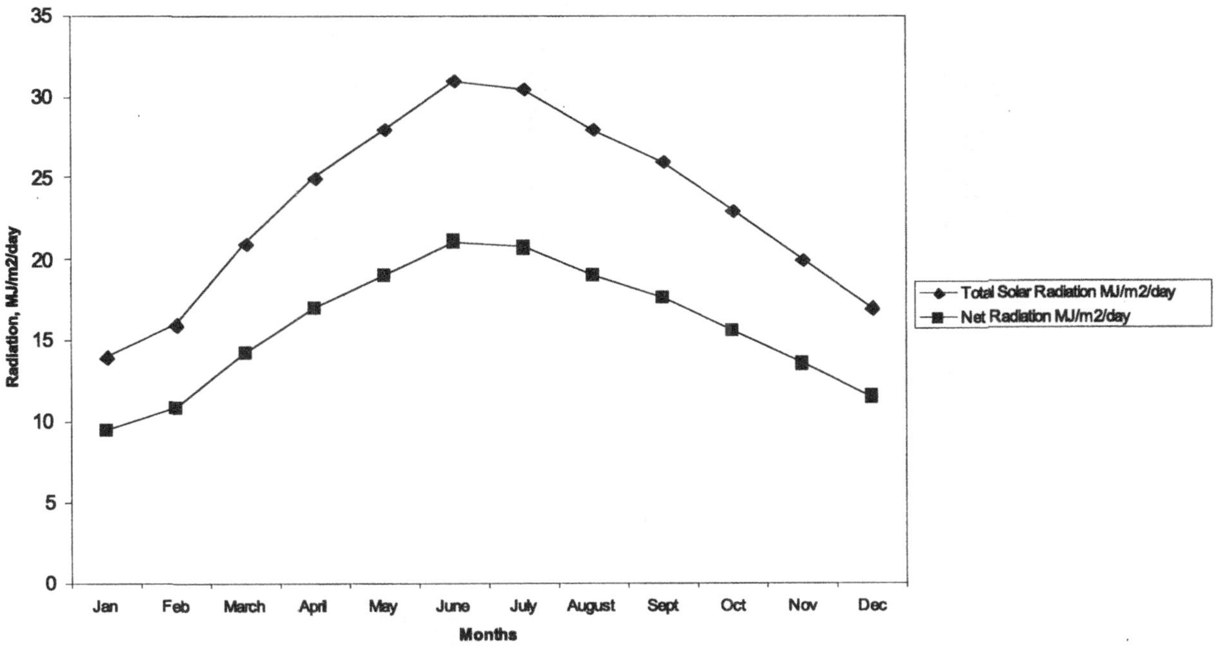

Figure 1.1.1: Monthly average of daily global and net solar radiation

The average available monthly mechanical energy over the peak months of May to August is presented in Figure 1.1.2 with an overall average of 58.6MJ. It is this characteristic that makes Alexandria a favourable site for solar water pumping (Hasson and Peck, 2004). The monthly average daily amount of water that can be pumped from a depth of 14m is shown in Figure 1.1.3. Daily pumped volumes of over 760m^3/day were achieved.

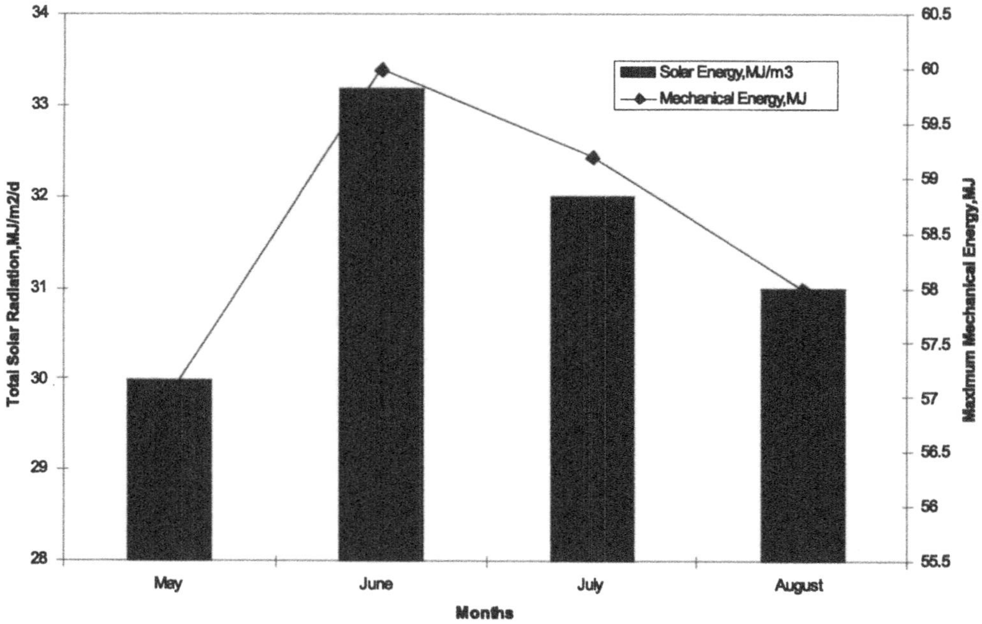

Figure 1.1.2: Monthly average of daily total solar radiation and maximum mechanical energy at the experiment site.

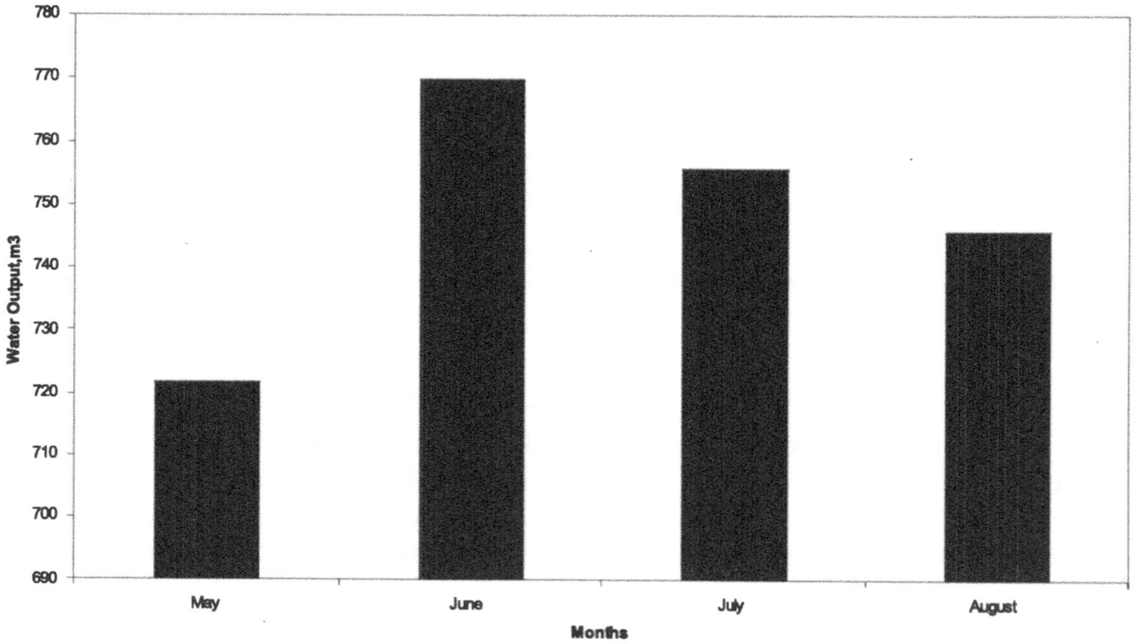

Figure 1.1.3: Monthly average of daily water output at the experiment site.

The solar radiation at Alexandria bears strong similarities to regions around 30° south in southwest Western Australia. For example, a monthly average daily radiation of between 11 to 32 $MJ/m^2/d$ in Iraq at latitude 30° north compared to between 10 to 30 $MJ/m^2/d$ between Moora to Katanning in Western Australia. From this predictions of solar pumping performance between Iraq and Western Australia can be made. Solar pumping closely matches the dry season water demand in both countries and is one of the inherent benefits of this arrangement (Hasson and Peck, 2004).

The potential evaporation rate was measured as described previously and the reduction in evaporative loss achieved by the palmfibre mat is shown in Figure 1.1.4. Typically a reduction of over 10mm/day compared to pan evaporation is achievable. In the peak critical months a reduction of nearly 20mm/day was achieved. The amount of short-wave solar radiation absorb by the mat and converted to kinetic energy depends on the mat surface reflectance (Kaisy and Hasson, 1992).

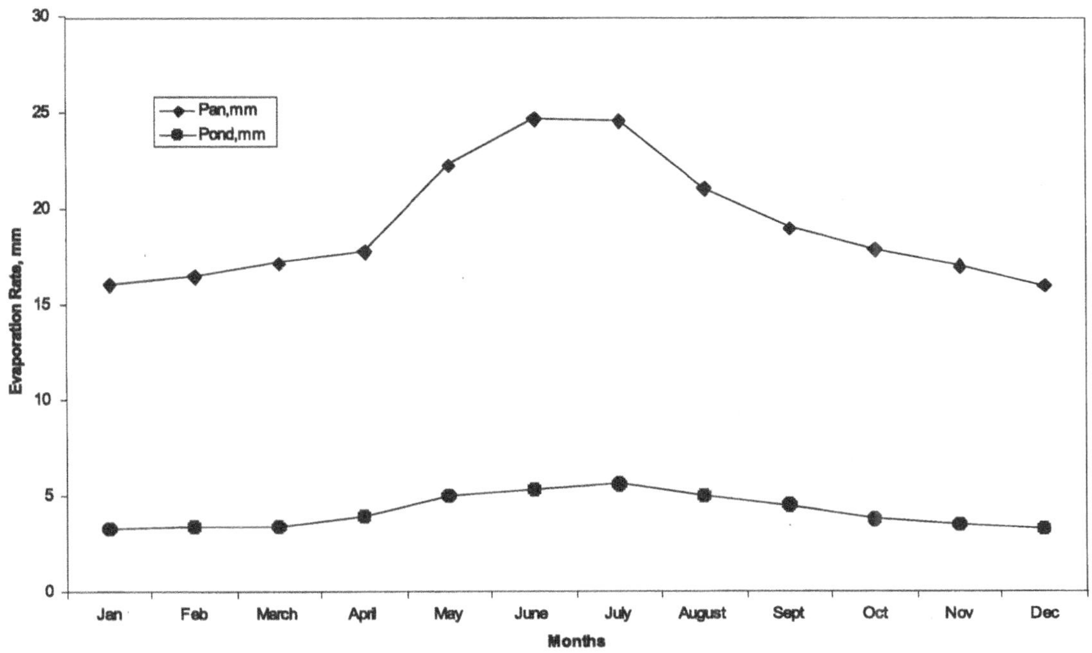

Figure 1.1.4: Monthly average of daily pan and pond evaporation rate at the experiment site.

The date-palm leaves mat can reduce the evaporation rate to 15 – 20 per cent of the net evaporation pan. Potential pan evaporation rates range from 17 – 24mm/day. The net evaporation rate from the pan located on the farm near the pond exceeded the rate from the pan in the middle of the pond due to the higher relative humidity.

The pumped water was stored in ponds to mix with rainfall water to be reused in irrigation of salt tolerant crops such as alfalfa, barley and used in stock drinking.
The pumped water was transferred by polyethylene lined ditch to the ponds over 300m distance at a slope of 2.25%. The pumped high salinity water mixes with the fresh rainfall water in the pond which is covered with date-palm matting to reduce the evaporative processes and blended water is finally transferred as brackish water.

Beside the water pumping technique there are other approaches that can be taken to reduce water evaporation such as decreasing the exposed water surface. Date-palm leaves matting however has been found to be the best method in terms of cost by which to cover the open ponds to minimize evaporation. Water losses of up to 50% by evaporation from open water storages in regions in Western Australian environment especially during the summer months can occur. Reducing this water loss using covers offers substantial economic benefit.

Conclusion

Solar energy is abundant in most of Iraq and Western Australia and the application of solar powered water pumping in Iraq shows considerable potential. Water scarcity and groundwater salinity are common problems in both these regions. In this research a maximum volume of $770m^3$/day using $27m^2$ of photovoltaic array was pumped from a depth of 14m. Subsequent dilution of this brackish water with rainfall runoff into ponds covered with floating palm frond matting achieved a final water quality suitable for agricultural irrigation purposes. This demonstrated that an integrated approach to solar use (via photovoltaics for energy) and control (via shading to reduce evaporation) in conjunction with water management could provide a viable economic solution.

In a forthcoming paper the Analytic Hierarchy Process (AHP) will be used for studying the feasibility of solar water pumping systems and reuse of discharged saline water and mixing with rainwater for agricultural applications.

References

AlSajir, B and Hasson, A.M., (1990). A procedure for minimizing of photovoltaic pumping system in irrigation applications in Baghdad. Solar & Wind Technology 7: 2.

Bilbeisi M., (1992). Jordan's water resources and the expected domestic demand by the yeas 2000 and 2010, detailed according to area. Proc. Symp. Jordan's Water Resources and Their Future Potential, P. 113-121

Gocht J. and Gillett B., (1985). Solar Water Pumping: a Handbook, Intermediate Technology Publications.

Gocht W., (1993) Economic feasibility of renewable energy systems. Proceedings of 4th Arab International Solar Energy Conference. P. 989-1001.

Hammad, M.A. (1999). Characteristics of solar water pumping, Energy; 24,85-92

Hasson, A. M. and AlSajir B., (1989). Relations of solar radiations and evaporative components in arid conditions. Science and Engineering Journal, 322-334.

Hasson, A. and Peck A., (2004). Solar powered groundwater pumping. Salinity Engineering Solutions Symp., Perth, Western Australia.

Leany and Christian., (2000). On-farm and community-scale salt disposal basin on the riverine plain: Evaluation basin leakage rate disposable capacity and plume development. CSIRO Land and Water Technical report 17/00.

Kaisy, A. and Hasson, A. (1990). Influence of soil moisture on soil temperature and heat storage. Journal of Agric. Engineering Res., 45:241-252.Effect of soil moisture.

Saaty T.L. (1980). The Analytic Hierarchy Process. New York: McGraw-Hill International.

Singh and Christian, (1999). On-farm and community-scale salt disposal basin on the riverine plain: Minimising the cost of basins, sitting design and construction factors. CSIRO Land and Water Technical Report 12/99.

1.2 Utilising An Integrated Wastewater Hydroponics System For Small Scale Use

Oyama, N*[1], Nair, J* and G.E. Ho*
*Environmental Technology Centre, Murdoch University, Western Australia, 6150
([1]Email: J.Nair@murdoch.edu.au)

Abstract

In areas where population increase is rapid and the centralised wastewater treatment system is unavailable, decentralised wastewater treatment systems are ideal. Due to the increase in demand for quality water and water scarcity world wide, reusing treated domestic effluent is a viable option. Most domestic wastewaters are high in the macro-nutrients and some micro-nutrients required for food production. Pathogen contamination is the major fear for using treated effluent to grow edible food crops, although there has been little evidence of this. A method of reducing the risk of pathogen contamination is to use the hydroponics technique. This ensures that the edible parts of the plants are not in contact with the wastewater. This study assesses the viability of using secondary treated wastewater from a decentralised treatment system using hydroponics to grow edible food crops and flowers compared to a commercial hydroponics medium.

Tomatoes and pansies were grown using secondary treated effluent in a hydroponics set up. Effluent analyses to test nutrient and microbial quality were conducted to study the efficiency of hydroponics in the improvement of wastewater quality. The results showed that wastewater hydroponics is a suitable method of treating wastewater and also for growing tomatoes. The pansies did not grow well in wastewater compared to commercial hydroponics medium. The study showed that the chemical and microbial quality of the effluent after hydroponics met the guidelines set by the World Health Organisation within 28 days.

Key Words

Decentralised systems; horticulture; hydroponics; wastewater reuse;

Introduction

Due to rapid urban growth throughout the world, local water shortages have become increasingly frequent even in humid regions (Riper and Geselbracht, 1997). The problem is compounded in many arid and semi arid countries, where water is becoming a scarce resource. Planners are forced to consider other sources of water, which may be used economically and effectively to promote further development (Al-Jayyousi, 2003). Water reclamation for urban use has begun to receive considerable attention around the world (Riper and Geselbracht, 1997). Wastewater reclamation and reuse provides an innovative and alternative option for agriculture (Haruvy, 1997; Lu and Leung, 2003).

The increase in food demand by the world's growing population outstrips the production of agricultural produce. The mobilisation of land and water resources is proceeding fast and water is often the most important factor for increasing agricultural production (Angelakis *et al.*, 1999). The most accessible fresh water resources are now almost entirely committed, leaving only alternative water resources for irrigation (Angelakis *et al.*, 1999). High concentrations of nutrients (mainly nitrogen and phosphorus) are found in reclaimed wastewater (Unkovich *et al.*, 2004) that can be utilised in agriculture.

There are concerns with wastewater reclamation and reuse due to public health risks (Unkovich *et al.*, 2004) and therefore close monitoring of the infrastructure and wastewater being reused is recommended (Lu and Leung, 2003). When using wastewater for irrigation some of the parameters to consider are the quality of treated effluent, amount of effluent being used, the probability of contamination of surface water bodies and groundwater sources (Department of Primary Industries Water and Environment, 2001), quality of produce and productivity of both crop and soil (Unkovich *et al.*, 2004).

The primary objective of any wastewater reuse project is to minimise or eliminate potential health risks. Pathogenic organisms are the greatest health concern in agricultural use of wastewater though few epidemiological studies have established definitive adverse health impacts attributable to the practice (Cooper, 1991). Of primary concern is the possible introduction of microbial pathogens into the groundwater by using reclaimed water for irrigation purposes (Wall *et al.*, 2002). The possible accumulation of some toxic elements in plants and the intake of toxic materials through consumption of these crops must be carefully assessed before it can be adopted (Sadovski *et al.*, 1978).

Most semi-arid and arid countries have limited water resources (Grobicki and Cohen, 1999), and for this reason, it is important to look at on-site wastewater treatment systems to assist in water conservation. The treatment units can be integrated with hydroponics, which is an intensive plant growth option that also ensures maximum use of water (Oyama *et al.*, 2005). The treated effluent may then be used for intensive horticulture to further utilise the remaining nutrients. It is essential to ensure that before reusing wastewater, the relevant guidelines are followed (Salgot and Pascual, 1996).

Hydroponics is a method of growing plants in an artificial nutrient medium with controlled physical and chemical conditions. Hydroponics is gaining increased attention as a bio-integrated food production system as it helps meet the goals of sustainability by following certain principles; the waste products of one system are recycled as food for a second biological system, increasing diversity and enhancing the system stability by regulating nutrient levels naturally (Diver, 2000). This study assesses the suitability of hydroponics for treatment of wastewater to meet the disposal quality set by World Health Organisation and the Australian Capital Territory (1999) Wastewater Reuse for Irrigation and compares plant growth using wastewater and hydroponics medium.

Materials and Methods

The experiment was conducted in 42L (59 x 35 x 23 cm) plastic storage containers. The wastewater (WW) was taken from the clarification chamber of the BioMax system, located at the Environmental Technology Centre (ETC), Murdoch University, which is an aerobic treatment unit that utilises both aerobic and anaerobic bacteria to breakdown blackwater and greywater. The hydroponics medium (Ag-Grow) used as control for the study was made according to the instructions to a ratio of 5mL of hydroponics commercial medium (CM) to 1L of water. This is a general-purpose hydroponics' plant nutrient medium.

Uniform sized seedlings of tomato and pansies bought from a nursery were used for the study. The experiment was conducted with 12 seedlings each of tomato and pansy in separate containers. There were a total of 4 hydroponics set ups, 2 for WW and 2 for CM. The experiment was conducted for 12 weeks in an indoor laboratory to minimise loss of water through evapotranspiration with adequate light. Water analysis and growth measurement of plants were conducted weekly. Water was changed depending on the volume reduction in medium level. The WW and CM samples were not filtered and were analysed for pH, nutrient analysis of ammonium-nitrogen, nitrate-nitrogen and phosphate were conducted using the HACH (2000) methods that are approved by the American Public Health Association (APHA). The microbial quality of WW was analysed by testing for faecal coliform and salmonella in 100mL (CFU/100mL) sample using membrane filtration method (Standards Australia, 1995a, b). The plant growth was measured by, noting the stem growth, counting the number of leaves, flowers and/or buds, fruits present and any colour change.

Results

Fresh medium was added on days 42 and 77, when it was found that the volume had reduced.

pH: The changes in pH levels are shown in figure 1.2.1. In the first 21 days, the pH decreased (6.08 –5.08) in all media and the trend is similar in the WW tubs and in the CM tubs. The next 25 days showed was a fluctuating pattern in all media. From the 56th day, the pH increased in all media except in the WW pansies.

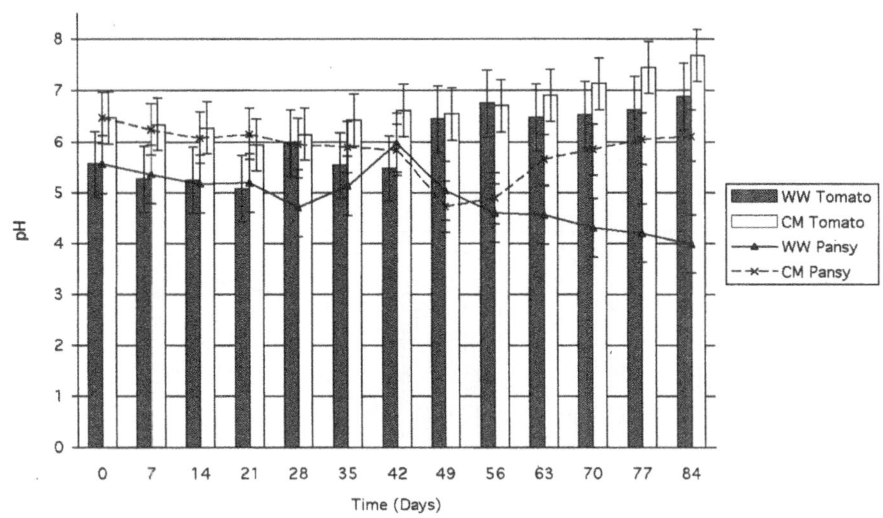

Figure 1.2.1: Changes in pH levels in WW and CM during plant growth. The error bars indicate standard deviation (WW Tomato ± 0.65; CM Tomato ± 0.51; WW Pansy ± 0.57; CM Pansy 0.5, n = 13).

Ammonium-Nitrogen: Figure 1.2.2 illustrates the increase in ammonium concentration in WW ± 1.04 (standard deviation), between days 14-21. By the 35th day, the ammonium concentration had dropped. The increase in the 42nd day is when new media was added. The ammonium concentration generally decreased in WW tomatoes, 0.005mg/L, CM tomatoes, 0.071mg/L and CM pansies 0.127mg/L by the 84th day when the plants were large, except the WW pansies (1.33mg/L).

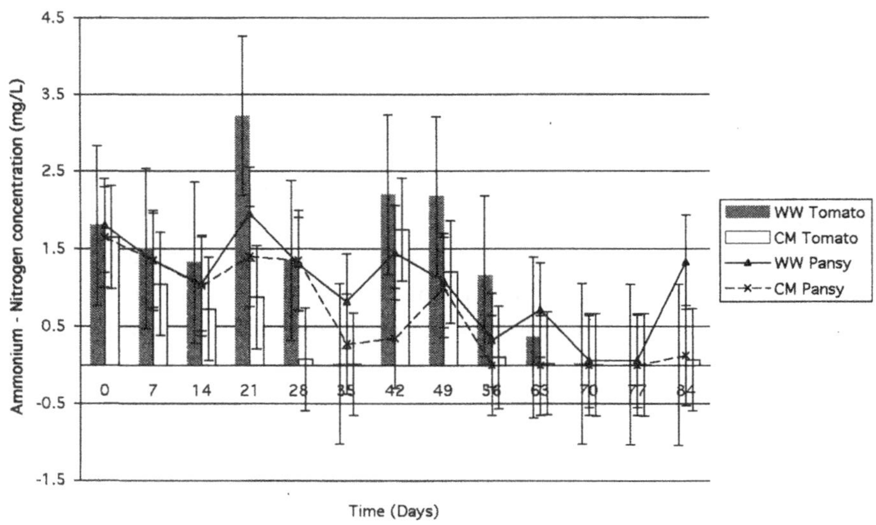

Figure 1.2.2: Changes in ammonium-nitrogen levels (mg/L)in WW and CM during plant growth. The error bars indicate standard deviation (WW Tomato ± 1.04; CM Tomato ± 0.66; WW Pansy ± 0.61; CM Pansy 0.65, n = 13).

Nitrate–nitrogen: The reduction in nitrate-nitrogen was slow initially when the plants were small until the 35th day. The increase noted for all media at day 42 and 63 occurred with the addition of fresh media. The drop was drastic during the final weeks for WW tomatoes (76%), CM tomatoes (81%), CM pansies (3.6%) while in WW pansies it increased by 17% as shown in figure 1.2.3.

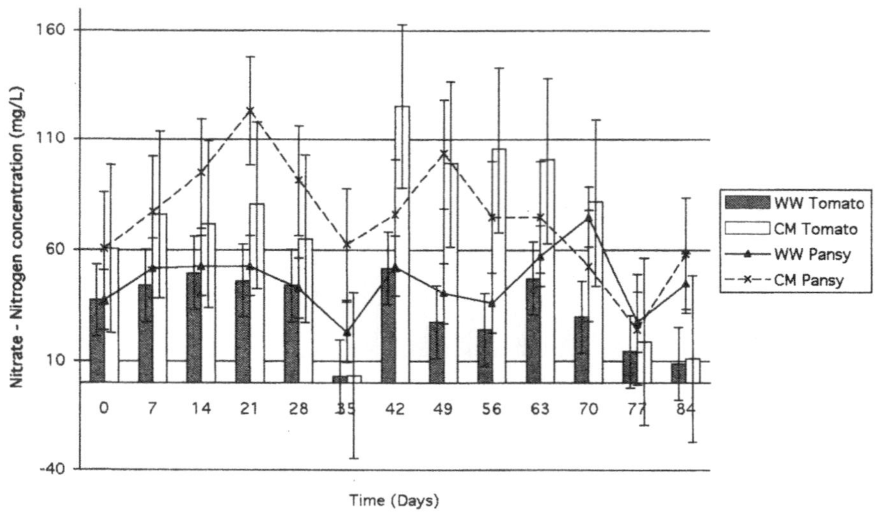

Figure 1.2.3: Changes in nitrate-nitrogen levels (mg/L) in WW and CM during plant growth. The bars indicate standard deviation (WW Tomato ± 16.36; CM Tomato ± 37.7; WW Pansy ± 13.57; CM Pansy 25.01, n = 13).

Phosphate-phosphorus: Phosphorus concentration decreased in the WW tubs from about 12.9mg/L in both WW tomatoes and pansies to 3.04mg/L for the first 35 days while the drop was not drastic in CM. After the 42nd day, the concentration fluctuated in all tubs as is shown in figure 1.2.4.

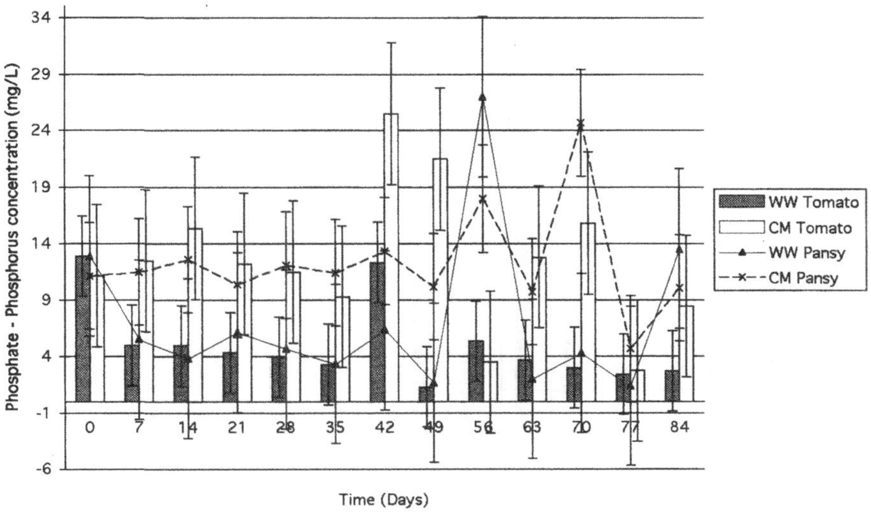

Figure 1.2.4: Changes in phosphate-phosphorus levels (mg/L) in WW and CM during plant growth. The error bars indicate standard deviation (WW Tomato ± 3.56; CM Tomato ± 6.29; WW Pansy ± 7.08; CM Pansy 4.73, n = 13).

Microbial quality: Table 1.2.1 shows die off of *E.coli* and *Salmonella sp* during the growth periods. It has been observed that there was a complete die off of the pathogens in 21 days.

Table 1.2.1: Colonies of E.coli and Salmonella sp. (cfu/100mL) in wastewater medium.

		Initial growth period			Second growth period			
		Day 7	Day 14	Day 21	Day 7	Day 14	Day 21	Day 28
Escherichia coli	Tomato	48	2	0	80	114	100	0
	Pansy	2	132	0	8	162	0	0
Salmonella sp.	Tomato	308	0	0	0	4	0	0
	Pansy	142	16	0	0	0	0	0

Plant growth (height, cm): There was not much difference in the growth (height) between the tomatoes grown in wastewater and hydroponics medium, while pansies grew better in CM than WW as shown in figure 1.2.5.

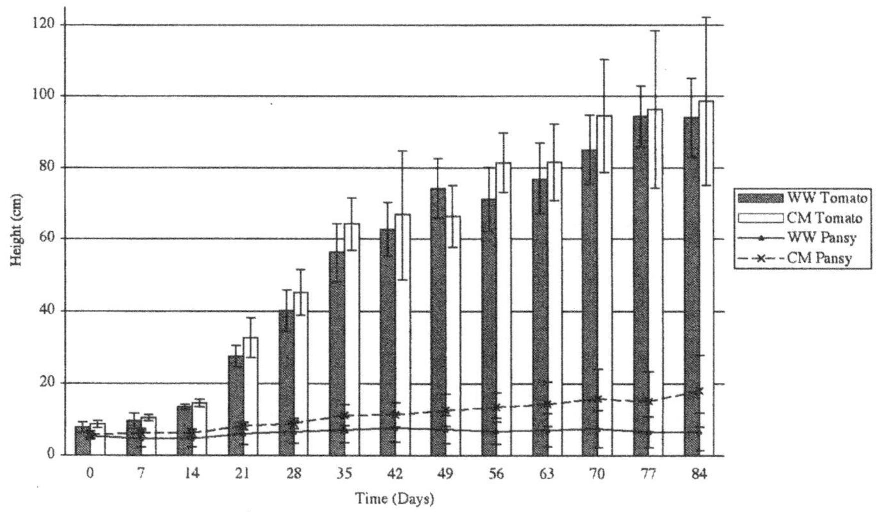

Figure 1.2.5: Comparison of height (cm) between plants grown in WW and CM.

Number of flowers: There were more flowers produced from the CM media. The error bars indicate standard deviation, $n = 12$.

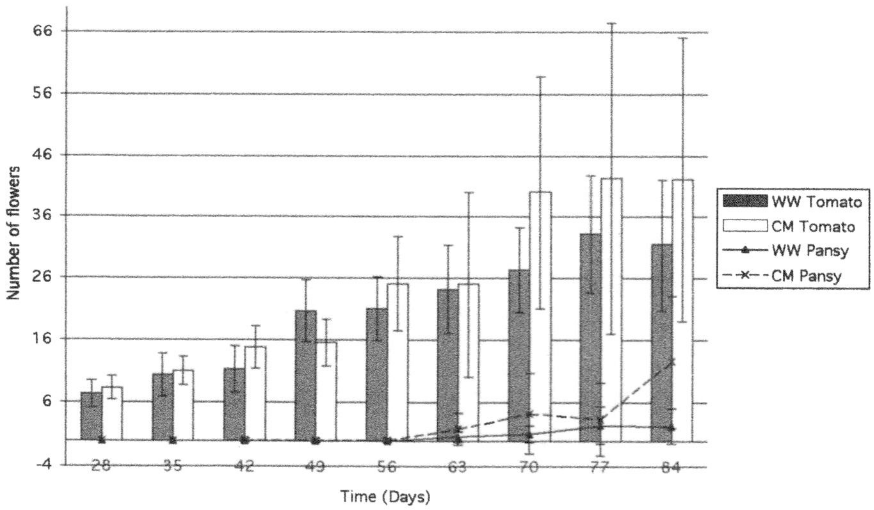

Figure 1.2.6: Comparison of number of flowers in WW and CM grown tomatoes and pansies. The error bars indicate standard deviation, $n = 12$.

Number of leaves and fruit: Overall, the tomatoes grown in WW and CM increased at approximately the same rate. With the pansies, the ones grown in the CM had a better growth rate as can be seen in table 1.2.2.

Table 1.2.2: Production of leaves and fruits with the total WW and CM used.

Day	Number of Leaves ± Standard deviation			
	Tomatoes		Pansies	
	WW	CM	WW	CM
0	13 ± 1	13 ± 2	12 ± 4	14 ± 5
7	26 ± 3	24 ± 2	9 ± 6	17 ± 5
14	46 ± 8	47 ± 4	16 ± 11	27 ± 7
	Number of Fruit ± Standard deviation			
	WW	CM		
70	0.4 ± 1	0.8 ± 1		
77	1 ± 2	2 ± 2		
84	4 ± 4	3 ± 3		

Total weight of fruit and total medium used: Overall, the plants grown in WW produced slightly more tomatoes than those grown in CM. Total amount of media used was the same for tomatoes (134L) and pansies (57L).

Table 1.2.3: Total fruit produced and the total WW and CM used.

Media	Tomatoes			Pansies
	Total ripe fruit (kg)	Total unripe fruit (kg)	Total media (L)	Total media (L)
WW	1.15	0.7	134	57
CM	1.05	0.6	134	57

Discussion

The present study showed that the nutrient levels (ammonium-nitrogen, nitrate-nitrogen and phosphate) in the wastewater decreased during the growth of plants. The reduction was more when the plants were larger. The decrease was more rapid in the WW than the CM.

The decrease in pH noted in all the media for the first 21 days (figure 1.2.1) when it began to stabilise, could be due to the fact that when the plants were smaller nitrification may not have occurred. The fluctuations between the 21st and 49th day possibly resulted from different flowering rates of the plants where they can take up varying amounts of nutrients, similar to the nitrification process (Pierce *et al.*, 1998). Mara and Cairncross (1989) recommend that the pH quality of wastewater for disposal between 6.5 and 8. The final pH in the WW tomato tub (6.9) and the CM tomato (7.7) complies with this regulation. However, the WW pansy tub had a lower pH (3.9), which can result in micronutrient toxicity (Whipker *et al.*, 2002). The CM pansy tub pH also reduced (6.1), which can result in boron or iron deficiencies (Whipker *et al.*, 2002). The reduction in pH could be due to the smaller rate of uptake shown by the stunted growth of plants or the decay of plants.

As plants are only able to take up nitrate-nitrogen, the conversion of ammonium-nitrogen to nitrate-nitrogen in the process of nitrification is essential. The maximum nitrate-nitrogen concentration (280.5mg/L) noted in figure 1.2.3 for the tomato CM relates to the low ammonium-nitrate concentration (0.025mg/L), shown in figure 1.2.2. The final nitrate-nitrogen concentration acceptable for disposal, or for further use for irrigation, is required to be between 10mg/L and 50mg/L (Australian Capital Territory Government, 1999), the levels in WW tubs (tomato tub was 8.8mg/L and in the pansy tub was 45.1mg/L), were within these guidelines. However, the guidelines set by Mara & Cairncorss (1989) require the concentration to be between 10mg/L and 30mg/L.

In order for plants to take up organic phosphorus, it is converted into mineral phosphorus by bacterial activity. The roots may take up particulate phosphorus by filtration (Noström *et al.*, 2003). There is an increase in the 42nd day (figure 1.2.4), which could be attributed to the addition of media in the tubs. From the 42nd day to the 84th day, large fluctuations in all the tubs except the WW tomatoes were noticed, possibly due to the release of phosphates within the particles suspended in the media (Chambers *et al.*, 1997). Increase in the phosphorus concentrations may also be due to a reduction in growth rate of the plants (Redding *et al.*, 1997), thereby reducing the amount of phosphorus uptake.

The highest reductions in nutrient levels occurred between days 0 - 35 in the tomatoes (ammonium-nitrogen-99%, nitrate-nitrogen-91% and phosphate-74 %). It is probably at this time the plants were starting their growth (figures 1.2.5 and 1.2.6) causing an increase in uptake to support growth. It is also at this time that the wastewater in the pansy tub experienced the largest reductions in nutrients (nitrate-nitrogen-38% and phosphate-74%) with the exception of ammonium-nitrogen, which occurred in days 42 - 54 (77%). This increase coincided with initial growth of the plant as well as flowering, which began from day 28 (figure 1.2.6). The CM tomatoes and pansies did not follow the same pattern as the WW. The reductions of nutrient occurred at different growth stages with CM tomatoes, ammonium-nitrogen between days 0 - 35 (99%), nitrate-nitrogen days 63 - 84 (96%) and phosphate, days 42 - 54 (86%). The reduction in CM pansy tub occurred between days 42 - 54 for ammonium-nitrogen (100%), 9 - 12 for nitrate-nitrogen (22%) and there was no reduction detected in those weeks in phosphate. The CM nutrient reductions may be due to the different requirements during the different growth periods of the plants.

Due to concerns to human health and safety regarding reusing wastewater bacteriological quality of the effluent is important. In this experiment a complete die off of *E. coli* and *Salmonella sp* occurred by the 21st day. Wall *et al.*, (2002) observed that the presence of indigenous groundwater micro-organisms significantly reduced pathogen survival time. The extent of exposure of the effluent to the edible part of the plant plays a big role in determining the types of food crops that can be grown using reclaimed wastewater (Christen *et al.*, 2006). The hydroponics system has very little exposure of the edible parts of the plant to wastewater, and as such reduces the risk of pathogen contamination. The growth of indigenous bacteria could have assisted in the pathogen die off in this system. According to Mara and Cairncross (1989), the recommended levels of colony forming units (cfu) should be 100cfu/100mL if the effluent is in direct contact with food. The final count for both bacteria satisfied this criterion in approximately 28 days, thus this water may be reused for purposes like irrigation as opposed to discharging it. Easa *et al.*, (1995) found that a retention period of 21 days is required, which ensures that the coliform bacteria count is within the guidelines set by World Health Organisation. The WW used in this experiment has undergone secondary treatment in the BioMax system and its use is more advantageous as it still contains nutrients that plants are able to take up.

Plant growth in WW and CM were similar. The total weights of tomatoes (Table 1.2.3) in both media were similar, indicating that the growth and production of flowers and fruits were comparable. The observations on the growth showed that there was a reduction in growth rate of all WW pansies and yellowing of leaves was observed. The most likely scenario is that the WW was lacking in micro-nutrients such as iron, which may be required for flowering and/or fruit production. Farm note (1984) describes iron deficiency, as yellowing of leaves and this may be the reason for the discolouration that was observed. However, yellow leaves could also be a symptom of nitrogen deficiency (Weir and Cresswell, 1993).

Conclusions

Crop production using secondary treated effluent (wastewater hydroponics) is a viable option as this ensures that the nutrients are utilised. In remote arid communities, this is especially important as the community is reusing their wastewater for crop production and the effluent from this system can be used for further irrigation for maximum conservation. Since the growth (height) of pansies during the first five weeks were similar to that grown in commercial medium, there is a potential for using treated wastewater for nursery use. From 134 L of WW, 1.15kg of tomatoes were produced which illustrates the potential for using hydroponics in conjunction with an on-site treatment systems in arid and remote communities.

The study showed that the chemical and microbial quality of the effluent after hydroponics met the guidelines set by the NHMRC (2000) within 28 days. The production of the tomato fruits in both media was similar. The growth of the tomato plants was much better than the pansies in the wastewater. This may be due to lack of micro-nutrients and or macro-nutrients as well as the size of plants. Research needs to be conducted to determine the micro-nutrient levels in wastewater and in the selection of crops that can be grown in wastewater hydroponics system.

References

Al-Jayyousi, O. R. (2003) Greywater reuse: towards sustainable water management. *Desalination*. ***156***, 181-192.

Angelakis, A. N., Marecos do Monte, M. H. F., Bontoux, L. and Asano, T. (1999) The status of wastewater reuse practice in the Mediterranean basin: need for guidelines. *Water Research*. **33** (10), 2201 – 2217.

Australian Capital Territory Government. (1999) *Act Wastewater Reuse for Irrigation: Environmental Protection Policy*, Canberra, Australia. Environmental ACT BDM99/0415.

Chambers, L., Olley, J., Crockford, H. and Murray, A. (1997) Conditions affecting the availability and release of phosphorus from sediments in the Maude Weir pool on the Murrambidgee River, New South Wales. *Managing algal blooms: outcomes from CSIRO's multidivisional blue-green algal program*, Davis. J. R. (ed). CSIRO Land and Water, Canberra ACT, 41 – 50.

Christen, E., Ayars, J., Hornbuckle, J. and Biswas, T. (2006) Design and management of reclaimed water irrigation systems. *Growing crops with reclaimed wastewater*, Stevens, D. (ed). CSIRO Publishing, Victoria Australia, 107 – 124.

Cooper, R. C. (1991) Public Health Concerns in Wastewater Reuse. *Wat. Sci. Tech.* **24** (9), 55-65

Department of Primary Industries, Water & Environment. (2001) Emission Limit Guidelines for Sewage Treatment Plants That Discharge Pollutants into Fresh and Marine Waters.

Diver, S. (2000) Aquaponics – Integration of hydroponics with aquaculture. NCAT Agriculture Specialist. February.

Easa, M., Shereif, M., Shaaben, A. and Mancy, K. (1995) Public Health Implications of Wastewater Reuse for Fish Production. *Water Science Technology*. **32** (11), 145 – 152.

Farm note. (1984) *Hydroponics – plants without soil*. Western Australian Department of Agriculture. No. 128/84, Agdex 200/16.

Grobicki, A. M. W. and Cohen, B. (1999). A flow balance approach to scenarios for water reclamation. *Water SA*. **25** (4), 473 – 482.

Haruvy, N. (1997) Agricultural reuse of wastewater; nation-wide cost-benefit analysis. *Agriculture ecosystems and environment*. ***66***, 113-119.

Lu, W. and Leung, A.Y.T. (2003) A preliminary study on potential of developing shower/laundry wastewater reclamation and reuse system. *Chemosphere*. **52**, 1451 – 1459

Mara, D. and Cairncross, S (1989) *Guidelines for the safe use of wastewater and excreta in agriculture and aquaculture: measures for public health protection*. Geneva, World Health Organisation.

Oyama,. N., Nair, J. and Ho, G.E. (2005) Recycling of Treated Domestic Effluent from an On-site Wastewater Treatment System for Hydroponics. *Water Science and Technology*, ***51***(10), 211-219.

Noström, A., Larsdotter, K., Gumaelius, L., la Cour Jansen, J. and Dalhammer, G. (2003) A small scale hydroponics wastewater treatment system under Swedish conditions. *Water Science and Technology*. ***48*** (11-12), 161 – 167.

Pierce, J., Weiner, R. and Vesilind, P. (1998) *Environmental Pollution and Control 4th Edition*. Buttersworth – Heinemann, New York.

Redding, T., Todd, S. and Midlen, A. (1997) The Treatment of Aquaculture Wastewaters – A Botanical Approach. *Journal of Environmental Management*. **50**, 283 – 299.

Riper, C. and Geselbracht, J. (1997) Water reclamation and reuse. *Water Environment Research*. ***69*** (4), 554 – 565.

Sadovski, A.Y., Fattal, B., Goldberg, D., Katzemelson, E. and Shuval, H. I. (1978) High levels of Microbial Contamination of Vegetables Irrigated with Wastewater by Drip Method. *Applied and Environmental Microbiology*. **36** (6).

Salgot, M. and Pascual, A. (1996) Existing Guidelines and Regulations in Spain on Wastewater Reclamation and Reuse. *Water Science and Technology*, **34**(11), 261 – 267.

Standards Australia. (1995a) *Water microbiology: Method 7: Thermotolerant coliforms and Escherichia coli – Membrane filtration method*. AS 4276.7.

Standards Australia. (1995b) *Water microbiology: Method 14: Salmonellae* AS 4276.14.

Unkovich, M., Stevens, D., Ying, GG. and Kelly, J. (2004) Impacts on crop quality from irrigation with water reclaimed from sewage. Final Report (1C) Prepared for Australian Water Conservation and Reuse Research Program.

Wall, K., Toze, S. and O'Hara, G. (2002) The Fate of Enteric Pathogens in Artificial Recharge Schemes. *Proc of the International Conference on Sustainability of Water Resources, Murdoch University*. Section 2 Water Quality, No. 3.

Weir, R.G. and Cresswell, G.C. (1993) Plant Nutrient Disorders 3: Vegetable Crops. Reed International Books, Australia Pty. Ltd. New South Wales.

Whipker, B. E., Cavins, T.J. and Gibson, J.L. (2002) Managing Fall Pansy Fertilisation. *NC State University Floriculture Research,* July.

1.3 Sub-surface Dripline and Turf: A Wastewater Irrigation Case Study

Katherine S. Taylor[1], Martin Anda[1], and Jeffrey Sturman[1].
*Environmental Technology Centre, Murdoch University, Western Australia, 6150

Abstract

Utilisation of treated wastewater is becoming more important in south-west Western Australia (WA) as the population increases and potable water sources diminish. The effectiveness of subsurface dripline irrigation (SDI) for applying treated wastewater was assessed by installing four products to the manufacturer's specifications in three 5m x 5m plots in the town of Northam, WA. These twelve plots were turfed, had sandy loam soil, and were monitored for two seasons (2004 to 2006). Turf managers were interviewed about the benefits, limitations, and appropriate management of SDI systems. The turf at the Northam site was healthy. No evidence of significant emitter clogging was found. Soil moisture sensor readings indicated that the plots were over-watered. This could be ameliorated by using a soil moisture sensor to control irrigation. Sometimes patches of saturated surface soil occurred. No evidence of vandalism was found. The greatest problems were leaks at joins and flush valves failing to turn off during operation. Poorly installed plots which leak are thought to increase the public health risk. The turf manager surveys also gave evidence that SDI has no detrimental effects on turf health, but participants reported fewer problems in their systems than occurred in the Northam system. Generally, SDI appeared to be effective for applying treated wastewater to turf in south- west Western Australian conditions.

Introduction

The State Water Strategy objective to beneficially reuse 20% of Western Australia's (WA) wastewater by 2012 reflects the importance of conserving water resources. Country areas in WA currently reuse about 40% of their treated wastewater, or 5.5% of the state's wastewater, predominately by sprinkler irrigation of sports fields, parks, school grounds and golf courses (Government of WA, 2003).

The town of Northam has used sprinklers to apply treated wastewater to turfed areas for over 20 years. A report on Northam's wastewater recommended that subsurface dripline irrigation (SDI) be used instead of sprinkler irrigation in order to reduce health risk (Sinclair Knight Mertz 1996), but using SDI for that purpose is uncommon in WA. Consequently, the WA Water Corporation funded a trial from 2004 to 2006 posing the question:

"Are currently available SDI products effective for the application of treated wastewater to turf?"

Literature Review: Subsurface dripline irrigation (SDI)

Subsurface dripline irrigation (SDI) consists of emitters placed at intervals within buried tubing. SDI is considered to be more water and energy efficient than sprinklers because it operates at lower pressures, and loses less water to evaporation, wind over-spray and run-off (Camp, 1998). SDI lowers the chance of human exposure to pathogens (Shuval *et al.* 1986). In comparative trials, SDI had less vandalism, maintenance (Gushiken 1995, Zoldoske *et al.* 1995), and plant disease than sprinklers (Solomon & Jorgenson 1993, Gratten *et al.* 1988). Turf irrigated with SDI can be of the same quality as sprinkler irrigated turf (Solomon & Jorgenson 1993, Zoldoske *et al.* 1995) when either treated wastewater or scheme water is applied.

Like any irrigation system, if SDI is incorrectly installed or poorly managed, or the water delivered does not match the soil's wetting characteristics, the irrigation efficiency decreases (Camp 1998). Although proponents of SDI claim that it has higher uniformity than sprinklers, an American study found there was little difference (Grattan *et al.* 1988).

Preventing emitter clogging is essential for SDI maintenance (Gushiken 1995, Solomon & Jorgenson 1993). Lines can clog with microbial films, inorganic debris (e.g. soil particles), or root intrusion. Each of these requires different preventative practices. Identifying and repairing damage from clogging or other means can be difficult and time consuming. SDI is also limited because it can't be used for establishing or germinating plants (Camp, 1998).

Coring and scarifying potentially pose a problem when SDI is used under turf, because the aeration equipment can damage shallow lines (Gilbert *et al.* 1981). A solution is to install lines deep enough to prevent damage. However, in sandy soils there may not be sufficient upward water movement if lines are installed "too deeply" (Bristlow *et al.* 2000), and the turf will become water stressed. The importance of aeration depends on the type of turf and intensity of wear (Beard 1972).

Materials and Methods

Four different SDI products supplied by were trialled over two irrigation seasons from November 2004 to May 2006. During the first season, the irrigation scheduling used the calculations detailed in Sturman *et al.* (2004). The scheduling was designed with the intention that each plot would receive an equal volume of treated wastewater. In the second season, irrigation was reduced. The volume of water applied during summer was about one third of what was applied in the first season. Repairs and maintenance, plot flow rate (water meters), leachate volume and nutrient loading (zero tension pan lysimeters), volumetric soil moisture (soil moisture sensor), turf condition, and meteorological data were monitored. All statistical analysis tested the null hypothesis to 95% confidence level. The trial site is summarised below:

- Disused tennis courts with amended sandy loam soil in full sun
- Couch turf (*Cynodon dactylon*)
- Four products, 3 driptubes and 1 driptape
- Three replicate plots per product =Twelve trial plots (5m by 5m area)
- Non-watered control
- SDI installed by the suppliers at 100mm depth
- The supporting infrastructure installed by the WA Water Corporation
- Mowing and fertilization by Town of Northam at the same frequency as the nearby sprinkler irrigated area.

Results

Repairs and maintenance

The repairs needed on the SDI plots are described in Table 1.3.1. The arbitrary end of the 2004-2005 season is when the pressure reducing valve (PRV) was adjusted in July 2005, which corrected problematic pressure fluctuations. Mowing and fertilization is not shown in Table 1.3.5, but was done by the Town of Northam at same frequency as the sprinkler irrigated area.

Table 1.3.1: Repairs and miscellaneous events at the Northam trial site. "Replacement lines" were put in at the end of the first season after some lines were removed and tested, but the joins on the new lines were not sound. Plot number (#) shown next to event.

Repairs 2004-2005	Repairs 2005-2006
Start: December 14th 2004.	September: Valve come off #1
December: Leaky air valves: #7, #11, #4.	September: Leak #5
December: Damage from backhoe: #8.	December: loose manifold connection #1
December: Loose manifold connection: #3, #6, and #12.	December: air release valve leaky- cleaned- plot #3
December: flush valves kept flushing: #9, #12	December: loose replacement plot #6
January: Loose manifold connection: #3, #12, #1.	December: loose manifold connection #4
February: Loose manifold connection #6, #1.	December: Lines burst #6,7
February : Flush valve keeps flushing: #11.	*December: Auto-flush valves on half the plots replaced with manual flush*
February: Flush valve blew out: #12	*February 8th.Driptape PRV installed*
April: Split pipe in sub-main to plot: #4	February: Line blew out #12
April: Two-spotted mite over site	February: loose replacement #8
Winter 2005. Run for 1 hour a week.	April: manifold line end cap blow out #4
July: PRV adjusted to stop pressure fluctuations	April : water in air release valve #11
July: loose manifold connection plot: #2	April: loose replacement line #12, 1
July: Loose replacement line: #2, 9	April: leak between meter box and manifold #2
August: flush valves kept flushing #9, 11	May: Loose replacement line #3
September: loose replacement #9, 3	May: loose manifold#12
	May: Leak #5, #9, probably replacements
	July 19th: Decommissioning begins

The average number of repairs needed over the two years, not including the commissioning period in December 2004 or the replacement lines, was 17.6 per year.

Soil moisture

The soil moisture varied between plots and depths. The variation in soil moisture could have been due to variation in the soil. The soil on site has been amended several times. Table 1.3.2 below compares the mean moisture of the SDI plots in the two seasons. Although it might be expected that the soil moisture would be less in the second season due to the reduced amount of water applied, it was significantly higher in all months, except June. The difference could be attributed to the higher evaporation rate in the first season. The only plots which did not have significantly higher moisture during the second season were plots #6 and 7.

Table 1.3.2: Mean soil moisture of SDI plots in selected months

Month	2004-2005 mean moisture	2005-2006 mean moisture	Statistical significance of mean difference
December	21.9	24.3	0.013
January	19.8	28.0	0.000
April	24.3	28.0	0.002
June	22.5	24.2	0.059

Soil salinity

The electrical conductivity (EC) of the upper (top 50 mm) layer of soil was significantly higher than the EC of the next layer (50 to 100mm) in the SDI plots and also the unwatered control. In the sprinkler sample, the EC was also higher in the top layer, but the difference was not statistically significant. EC in all plots increased over the season (Table 1.3.3).

Table 1.3.3: Soil saturated extract electrical conductivity (micro Siemens) in two layers of soil

	SDI plots		Unwatered		Sprinkler plots	
Sample depth	0-50mm	50-100mm	0-50mm	50-100mm	0-50mm	50-100mm
29.09.05	1.39	0.7	1.5	1.8	0.9	0.6
08.02.06	1.43	1.05	2.51	1.49	1.23	0.97
06.04.06	2.98	2.09	3.24	1.2	3.12	1.71
25.05.06	3.25	2.06	3.3	1.26	1.7	1.18

Plot flow rates and water consumption

The gross water consumption of each plot during the second season is shown in Table 1.3.4 below. The average water consumption was 75kL, but there was variation. Plot 1 consumed over twice the average (192 kL), due to a series of leaks. In contrast, plot #2 used about half the average (39kL), possibly because it was turned off twice for repairs, and did not lose large volumes of water from these leaks.

Table 1.3.4: Plot water consumption in the second season (2005-2006)

Plot	1	2	3	4	5	6	7	8	9	10	11	12
kL	192	39	118	104	61	51	42	66	61	58	56	55

The flow rate data from 2004-2005 was unreliable. In the second season a more accurate method of measuring flow rates to plots was used. Only the data from 2005-2006 will be discussed here.

The flow rate to each plot varied over the season depending on the presence of leaks, although some of the variation in flow rates could be attributed to experimental error (i.e. error in reading water meters). The flow rates of plot 12 are fairly typical and shown in figure 1.3.1.

The average flow rate to each plot was higher than the manufacturer's specifications for all plots except plot #2, which was lower (Table 1.3.5). This was attributed to the leaks in the plots. The difference between plot flow rates and expected flow rates varied in plots with the same product. For example, plots with "driptube 2" had flow rates ranging from 89% to 204% of the flow rate expected from the manufacturer's specifications.

Table 1.3.5: Average flow rates of each plot in 2005-2006 compared with the flow rate expected from the manufacturer's specifications and expressed as a percentage.

Plot number	Expected flow rate (L/m^2/hr)	Average flow rate (L/m^2/hr)	% of expected flow rate
Driptape			
6	13	13.6	105%
7	13	15.0	115%
11	13	18.6	143%
Driptube 1			
4	11.85	18.1	152%
9	11.85	18.5	156%
10	11.85	23.7	200%
Driptube 2			
2	10	8.9	89%
5	10	20.4	204%
8	10	13.2	132%
Driptube 3			
1	10	14.9	149%
3	10	21.1	211%
12	10	15.3	153%

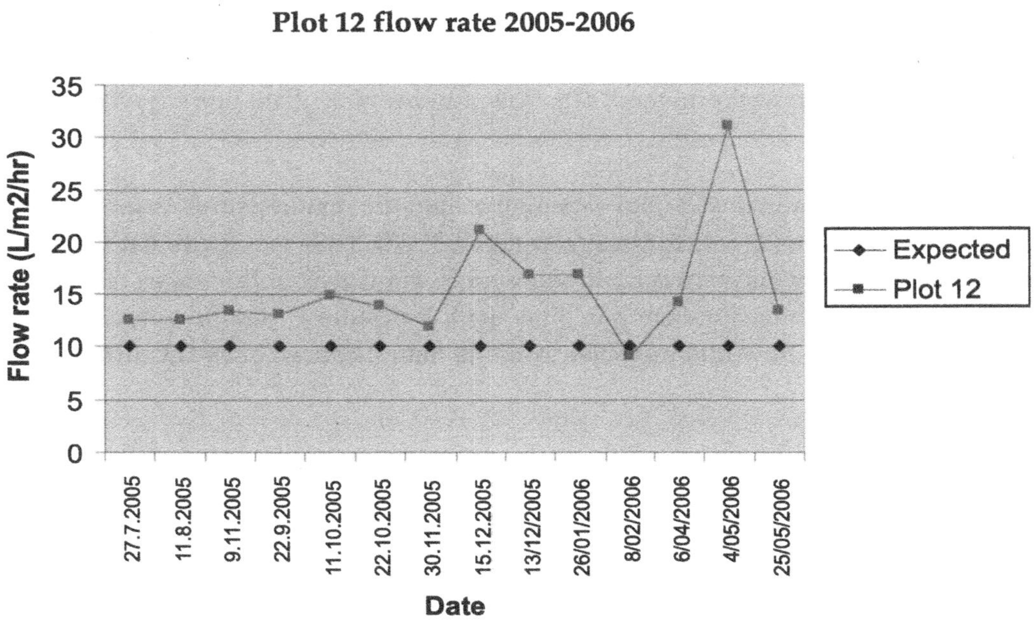

Figure 1.3.1: Variation in the flow rate of plot 12 compared to the flow rate expected from the manufacturer's specifications

Turf health

Although the turf took several months to establish in the first season, the cover was almost complete in the second season. The turf was quite thatchy compared to the sprinkler irrigated plots. An outbreak of two-spotted mite (*Tetranychus urticae*) occurred in the first season, but was not repeated in the second season.

Survey results

A total of 11 Australian turf managers who used SDI were interviewed. 4 used treated wastewater, and 7 used other water. In spite of being unable to use turf aeration equipment, most (63.6%) said they had no problems with the health of the turf. Some reported that it was slow to establish (27.3%), but only one participant (9.1%) said that the cover was patchy. 90.9% said either that using SDI was the same as using sprinklers or said that it was better.

Repairs were mainly (54.5%) damage to the lines from a sharp object (e.g. spade / stakes) being poked through it, which could be repaired in an average of half an hour. Two participants said there were no problems requiring repairs (18%), and only one (9.1%) reported leaks from product failure (e.g. joins coming undone). The only vandalism described were car "burn-outs" in one system (9.1%)

Discussion

The Northam Experience

The products performed similarly, except the driptape, which burst from the high pressure used. Installing four products in one system necessitated a compromise in operating pressure, and the failure of the driptape was the most tangible evidence of strain on the products.

Although less water was irrigated in the second season than the first, the soil moisture was higher. This was probably due to unusually high rainfall over the first few months of summer in the second season. The irrigation would have been more efficient with the use of a soil moisture sensor to control irrigation, (as described in Byrne 2002), especially because the soil had significant clay content. Most plots had unexpectedly high flow rates, which could also be compensated for by using a soil moisture controller.

The plot flow rates varied considerably, and so did the volume of water consumed by each plot. The differences are probably due to the presence of leaks. This illustrates that SDI is not an inherently water efficient technology. If there are problems in the system SDI can be very inefficient.

Leaks are of concern when treated wastewater is used because they can allow humans to be exposed to pathogens. Saturated soil patches can occur over shallow lines as well as leaks. In practice however, leaks may present low health risk. Shuval &Fattal (2003) reviewed over 1000 documents on wastewater irrigation, and found that there was no demonstrated risk to the general public from treated wastewater irrigation with sprinklers. They assert that "... reduction in coliforms to the level of a few thousand/ 100mL, but not total removal, can provide a high level of health protection" (p248.). If this is accurate, then the health risk of irrigating wastewater of Northam's quality (less than 10 thermotolerant coliforms per 100mL) to turf is effectively zero, regardless of whether SDI or sprinklers are used.

The results from the soil EC tests indicate that salt build up is not necessarily caused by using SDI. The high EC in the unwatered control indicates that salt build up is caused more by the absence of above ground flushing (e.g. by sprinklers or rain) rather than by the SDI itself. These results do not contradict the findings of Suarez-Rey *et al.* (2000) that salt accumulation near the soil surface is correlated with SDI usage. Rather, they contradict the conclusion that the accumulation is directly caused by SDI use. Surarez-Rey *et al.* (2000) also noted that there was insufficient accumulation of salt to affect turf health, which is supported by the results of the Northam trial.

The broader experience

The turf managers surveyed were generally happy with the performance of their SDI systems, regardless of whether or not they used treated wastewater. They seemed to have very few problems with turf health, which supports the findings of the Northam experiment. In contrast with the Northam experiment, the other systems had fewer problems with leaks. The majority experienced leaks as a result of carelessness (e.g. stakes driven through the ground), whereas at Northam the leaks were, presumably, from poor product installation or product failure. Although the number of leaks in the Northam system was undesirable, it appears to be atypical of SDI systems.

Conclusions

SDI is effective for applying treated wastewater to turf. It has the advantages of putatively decreasing human health risk, potentially increasing irrigation efficiency, and reducing vandalism. The Northam experimental system had a lot of problems, predominantly leaks, and one product (the driptape) failed due to using a high operating pressure. In contrast, the systems described in the survey did not have these problems. The other systems generally had low maintenance and did not need repairs often. Virtually no vandalism was experienced, which saved money and time. Most participants surveyed were satisfied with their SDI systems.

Yet there are disadvantages to using SDI. Aeration and turf renovation equipment can't be used. Using stakes, star pickets or other sharp objects in the ground is limited because it can easily damage the driplines. Installation and design (E.g. type of filtration, location of valves and operating pressure) are important aspects of effective systems to prevent on going problems as were experienced in Northam. While these potential problems are not insurmountable, they need to be considered when evaluating the effectiveness of SDI.

References

Beard, J.B., 1972. *Turfgrass: Science and Culture.* Prentice Hall, N.J., USA

Bristlow, K.L, Cote, C.M., Thornburn, P.J., and Cook, F.J., 2000. Soil Wetting and solute transport in trickle irrigation systems. In: *Proceedings of the 6th International Micro-irrigation Congress, Micro-irrigation Technology for Developing Agriculture.* South Africa, 22-27 October 2000.

Byrne, J., 2003. *An assessment of a low cost soil moisture controller for water conservation in domestic irrigation systems.* Honours Thesis. School of Environmental Science, Murdoch University.

Camp, C.R., 1998. Sub-surface drip irrigation: A review. In: *Transactions of the ASAE,* (41) 5. 1353-1365

Gilbert, R.G., Nakayama, F.S., Bucks, D.A., French, O.F., and Adamson, K.C., 1981 . Trickle irrigation. Emitter clogging and flow problems. In: *Agric. Water Manage.,* 3: 159-178

Government of Western Australia, *Securing our Water Future: A State Water Strategy for Western Australia.* State Government of Western Australia. Western Australia.2003

Grattan , S., Schwankl, L., and Lainin, W., 1988. Weed control by sub-surface drip irrigation." In: *Calif. Agric.,* 42 (3) pp22-24.

Gushikin, E.C., 1995. Irrigating with reclaimed water through permanent subsurface drip irrigation systems. . In: *Proceedings of 5th International Microirrigation Congress,* April 2-6, 1995, Orlando, Florida. American Society of Agricultural Engineers, Michigan pp 775-780.

Shuval, H., Adin, A., Fattal, B., Rawitz, E., and Yetutiel, P., 1986. *Wastewater Irrigation in Developing Countries. Health Effects and Technical Solutions.* 1 ed. Integrated Resource Recovery Series. Vol. 6. 1986, Washington, USA: The World Bank.

Shuval and Fattal, B., 2003. Control of pathogenic micro-organisms in wastewater recycling and reuse in agriculture. In: Mara, D., and Horan, N., 2003. *The handbook of water and wastewater microbiology.* Academic Press, London, UK.

Sinclair Knight Mertz (1996) *Northam Effluent Reuse Study FINAL REPORT.* Sinclair Knight Merz Pty Ltd.

Solomon, K.H., and Jorgensen, G., 1993. Research Report- Subsurface drip irrigation. CATI Publication # 930405. Available online: www.cati.csufresno.edu/cit/rese/93/930405/index.html.

Sturman, J., Ho, G., and Mathew, K., 2004. *Water auditing and water conservation.* IWA Publishing, London.

Suarez-Rey, E., Choi, C. Y., Waller, P. M., Kopec, D. M., 2000. Comparison of subsurface drip irrigation and sprinkler irrigation for Bermuda grass turf in Arizona. *Transactions of the Asae,* 2000. 43(3): p. 631-640.

Zoldoske, D., Genito, S. and Jorgensen, G., 1995. Subsurface Drip Irrigation (SDI) on Turfgrass: A University Experience. In: *Proceedings of 5th International Microirrigation Congress,* April 2-6, 1995, Orlando, Florida. American Society of Agricultural Engineers, Michigan pp 314- 324.

2 Small Wastewater Systems

2.1 Large On-site Wastewater Treatment and Reuse Systems for Commercial Applications

Ben Kele*, David J Midmore*, Barry Hood*, Brendan J. McKennariey**
* Plant Sciences Group, Central Queensland University,
Rockhampton Qld 4702 Australia
(E-mail: b.kele@cqu.edu.au, d.midmore@cqu.edu.au, b.hood@cqu.edu.au)

** GBG Project Management P/L P.O. Box 163, Qld 4054 Australia
(E-mail: brendan@gbgprojects.com.au)

Abstract

The Central Queensland University (CQU) in partnership with GBG Project Management has installed large on-site treatment and reuse for a variety of commercial operations in Queensland. Systems range in size from 30 equivalent persons (EP) to 500 EP. Installations include taverns, motels, child care centres, and schools. The effluent produced at large commercial sites is significantly different to domestic wastewater with respect to constituents and generation patterns. Standard on-site treatment technologies used at large commercial sites may need to be adjusted to ensure water quality performance. At the sites study treatment is obtained through the use of a combination of septic tanks, filters, disinfection technologies, and the KEWT system, a recirculating evapotranspiration trench. A variety of filter media is used, such as zeolite, scoria, charcoal, rock, and sand. The KEWT trenches contain filter media, soil, and selected plant species. Effluent passing through the KEWT trenches is filtered physically and biologically, prior to being transpired by the plants. Any excess effluent can be recirculated through the trenches or diverted for water recycling. Treated effluent at the sites is available for reuse purposes, predominately irrigation. Effluent flow through the treatment and recycling systems is regulated by remotely controlled SCADA technology. The SCADA system is an intelligent control device that automatically regulates flow throughout the treatment process minimising the risk of hydraulic surges.

Key Words

BOD_5, equivalent person, filter media, KEWT, SCADA, wastewater characteristics

Introduction

Commercial sized on-site wastewater treatment plants typically deal with a greater volume of wastewater than a domestic system. In addition to this the wastewater characteristics may not be within the typical domestic range and may comprise one or/more constituents that inhibit effective treatment (McAvoy 2002; Asquith, Shelly *et al.* 2005; Sostar-Turk, Petrinic *et al.* 2005). If the ratio of blackwater to greywater differs significantly this can have a major impact on biochemical oxygen demand (BOD_5), total nitrogen (TN), total phosphorous (TP), total suspended solids (TSS), dissolved oxygen (DO) and faecal coliforms (Tchobanoglous, Burton *et al.* 2003). Specialised types of wastewater, such as tradewaste or stormwater can also be diverted into commercial on-site wastewater treatment plants. Domestic on-site systems sometimes fail when they are upsized to treat commercial wastewater as the concentrations of the constituents are too dissimilar for comparable treatment performance (Tchobanoglous and Burton 1991). Asquith, Shelly *et al.* (2005) reported that commercial systems are frequently poorly understood by owners, maintenance personal, and regulators resulting in less than optimum performance with associated risks to public and environmental health from insufficiently-treated effluent. Asquith, Shelly *et al.* (2005) stated that the main problems were associated with design, installation and operation and maintenance.

Design issues centre on the variable flow often experienced in commercial systems due to the wastewater generation patterns. Commercial on-site systems at sites such as tourist facilities, schools, licenced premises, and child care centres rarely have typical domestic wastewater generation patterns due to events such as school holidays and closure at nights or on weekends. Alternatively peak use may occur at night or weekends, and at special events (Kele, Midmore *et al.* 2003; Asquith, Shelly *et al.* 2005). These types of wastewater production patterns can result in hydraulic surges in undersized systems, that in turn cause poorly treated effluent to be released into the environment (Panswad and Komolmethee 1997). Systems on licenced premises, such as pubs and taverns, can often be undersized as water flow assumptions typically used to design on-site wastewater systems can sometimes fail to account for the 'brewers factor'; which is the volume of liquid (not linked to the water-main) sold within the premises and excreted before the patron leaves the venue. Licenced premises can also have additional treatment issues due to the predominance of blackwater (especially urine) over greywater, the presence of alcohol by-products harmful to beneficial bacteria, and a relatively low concentration of biologically available carbon required for microorganism nutrition (Fleeger, Carman *et al.* 2003; Tchobanoglous, Burton *et al.* 2003). Other common design flaws include over- or under-sized land application areas and inappropriate combinations of treatment processes (Asquith, Shelly *et al.* 2005). Installation flaws include the selection of unsuitable components such as pumps, filter media, control equipment, alarms, as well as poor project management, resulting in the system not been installed as per plan (Asquith, Shelly *et al.* 2005). The main operation and maintenance issue is the failure to budget, financially and human resource wise, for the ongoing costs and time associated with running the treatment system. Other problems include changes to the commercial operation over time that result in alterations in wastewater characteristics and generation patterns, lack of understanding from the owner as to how the technology works, unauthorised alteration to the controls, and insufficient monitoring (Asquith, Shelly *et al.* 2005).

The Central Queensland University has for the past seven years researched a technique of on-site wastewater treatment and reuse suitable for use in commercial situations (Kele 2005). The technology is based on a version of recirculatory evapotranspiration channels (see Figure 1). Primary treatment is achieved through the use of anaerobic sludge chambers and in-line filters. Secondary treatment techniques use a combination of aeration, aggregate biofilms, volcanic rock filters, and sand, soil and rhizofiltration. Aspects of tertiary treatment are obtained from the biological activity of microorganisms and selected plant species as well as the cation exchange capacity of specific aggregates. Reuse is through evapotranspiration or if desired through a variety of other applications, such as irrigation. Non-transpired effluent is recirculated through the system until reused. The treatment and reuse method that has been developed is marketed as the Kele effluent wastewater treatment (KEWT) system. The technology is controlled and monitored through a Supervisory Control and Data Acquisition (SCADA) system that gathers information and controls on a supervisory level, and directs mechanical processes. It is purely a software package that is located with the hardware to which it is interfaced, in general via Programmable Logic Controllers (PLCs). The SCADA system has a computer program that will automatically adjust for hydraulic surges, as well as allow for remote monitoring of the system.

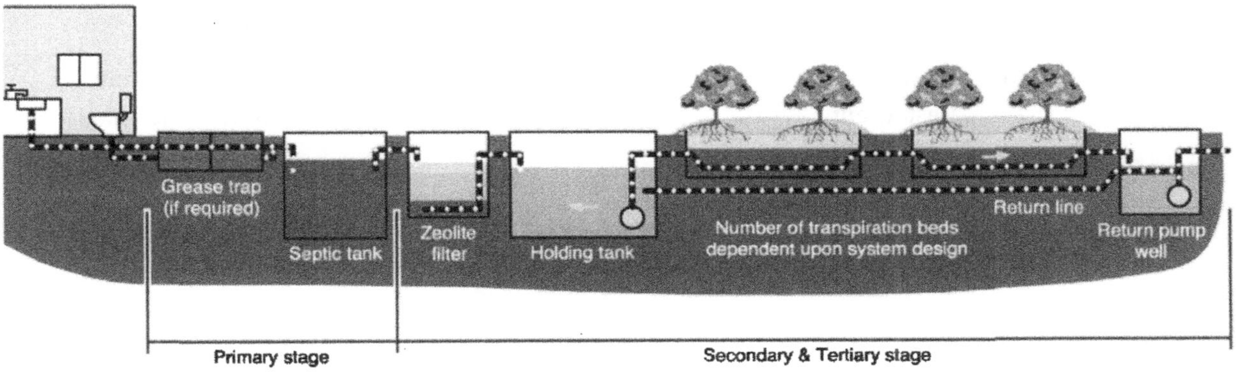

Figure 2.1.1: Schematic of the KEWT system

Methods

This paper will discuss six KEWT systems (see Table 2.1.1), located in central and south-eastern Queensland. The sites were installed according to the method described in Kele (2005). Installation methods at each site were altered to suit the specific requirements of the site; for example some sites were identified as carbon-source poor and had charcoal added to the filter media. The volume used per equivalent person (EP) was supplied by the relevant local government authority. The sites were all large commercial systems 70 EP or greater. The Burpengary site had a requirement that all the treated effluent was reused within the evapotranspiration section of the KEWT system. All other sites required recycled water from the KEWT system for other purposes. The Gem-Air site required recycled water for gem-mining and irrigation, the Comet site for sub-surface irrigation, the Pritchard Rd site for above-ground golf course irrigation, the Greenbank site sub-surface irrigation of a cabinet timber plantation and sports oval, and the Woodpecker site sub-surface irrigation. A specific SCADA program was written for each site to precisely cater for any unique characteristics. All SCADA programs report to a central monitoring location via redundant links of differing formats. These links consist of a physical link through the public switched telephone network (PSTN) and a redundant global systems mobile (GSM) link through the mobile network (MN). Both the PSTN modem and the GSM modem are powered via the uninterruptible power supply allowing site monitoring at all times. Depth sensors are installed in each tank and are PIL brand 50kHz 24 VDC ultrasonic transmitter/receivers with a 0 – 10 V analogue output proportional to the distance of the target from the sensor. Sensor range is 200 mm to 2000 mm. A monitoring program is linked to the alarm system and any incidents result in a message being sent to the central control station.

Table 2.1.1: Descriptions of the KEWT systems

Site	Function	EP	Installed	Recycled Water	Treatment Concerns
Gem-Air	Caravan Park	100	April 2002	Yes	Seasonal Flow
Comet	Pub and Miners Camp	120	June 2005	Yes	Alcohol by-products, laundry and shower wastewater from coal miners
Pritchard Rd	Motel, Restaurant and Miners Camp	150	Sept 2005	Yes	Alcohol by-products, laundry and shower wastewater from coal miners
Burpengary	Child Care Centre	70	Dec 2005	No	Mainly blackwater; predominately urine, no flow on weekends
Greenbank	Primary School	600	Feb 2006	Yes	Mainly blackwater; predominately urine, no flow on weekends or school holidays
Woodpecker	Tavern	140	March 2006	Yes	Alcohol by-products; mainly blackwater; predominately urine

All the KEWT sites discussed in this paper due to the size of the systems (>20 EP) require individual licences from the Queensland Environmental protection Agency (EPA). Each licence states the specific tests and the frequency required. Primary samples were collected from the outlet of septic tank. KEWT samples were collected either from the discharge point for recycled water or from the return line in the case of the Burpengary site which is a closed system. Samples were stored and transported at 4°C and were received at the laboratory within 24 hours. All analysis was conducted using NATA accredited laboratories using standard techniques from APHA (2001). Faecal coliforms were determined by Stockinette test MIMM 8.5, total combined nitrogen by method 208/240, total phosphorous by method 402 AES element by ICP, suspended solids by method 246, and biochemical oxygen demand 5 day by method 212 (APHA 2001).

Results

Table 2.1.2 shows average results from the Qld EPA licence compliance testing for the six selected KEWT systems It is clear that the quality of the primary treated effluent differed between the sites.

Table 2.1.2: Average Results from the EPA Licence Compliance Tests for the KEWT systems

Site	BOD_5 mg/L	TN mg/L	TP mg/L	TSS mg/L	DO mg/L	Faecal Coliform cfu/100 ml
Gem-Air Primary	**48**	**71**	**8**	**84**	**0.5**	**>3 000 000**
Gem-Air KEWT	7	2.4	0.6	9	4.6	2
Comet Primary	**235**	**52**	**23**	**230**	**0.2**	**2 560 000**
Comet KEWT	24	21	9.1	31	5.3	356
Pritchard Rd Primary	**134**	**81**	**42**	**46**	**N/A**	**1 450 000**
Pritchard Rd KEWT	7	23	11	24	N/A	28
Burpengary Primary	**198**	**70**	**6.4**	**220**	**0.9**	**1 125 000**
Burpengary KEWT	10	13	2	26	4.9	190
Greenbank Primary	**160**	**160**	**14**	**61**	**0.1**	**1 000 000**
Greenbank KEWT	<5	5.5	0.32	<5	6.1	560
Woodpecker Primary	**519**	**74**	**10**	**48**	**0.1**	**51 000**
Woodpecker KEWT	20	18	5	22	5.5	450

The Woodpecker site had the highest BOD_5 value; it is probable that this was due to it being a licenced premises with no accommodation. The Comet site also a licenced premises but with a 40 bed accommodation service attached had the next highest BOD_5 concentration. Greenbank and the Burpengary site both dealt with wastewater produced predominately by children, the relatively small volume of greywater produced at these sites is thought to account for the high BOD_5 readings. The Gem-Air site had a BOD_5 reading typical of domestic wastewater. The KEWT system has been able to lower the BOD_5 at all sites to within licence compliance values. The ability of the KEWT system to continue to do so at sites where the BOD_5 is relatively high (such as Woodpecker) is a matter for further research. The Greenbank site had the largest TN value most likely related to the relatively high volume of urine production. It is thought that the dilution factor of frequently running urinals lowered the TN at the licenced premises. The Burpengary, Gem-Air, and Pritchard Rd sites all had similar TN readings. The Gem-Air KEWT system had the lowest TN value, chiefly due to the fact that the plants in the evapotranspiration site are more mature and have a greater nitrogen nutrient requirement than the plants currently at the other sites. All KEWT sites complied with the TN licence requirements. The Pritchard Rd and Comet sites both had relatively high concentrations of TP which were linked to the types of cleaning chemicals used. The TP concentrations for Greenbank, Burpengary, Gem-Air, and Woodpecker were similar to typical domestic concentrations (Tchobanoglous, Burton *et al.* 2003). The Comet and Burpengary sites both had relatively large TSS values in the primary treated effluent and this was linked to poorly functioning in-line filters in the septic tank outlets. The filter at Comet site has difficulty removing the fine coal dust provided in abundance from the showers and laundry of the mining camp. All other sites had typical values for TSS (Tchobanoglous, Burton *et al.* 2003). The KEWT samples for TSS all meet with licence requirements; the tendency for segments of biofilm to break off the filter media during the recirculation of the non-transpired water results in suspended solids in the treated effluent. As expected the primary treated effluent had very low DO values; however the KEWT sample DO was higher showing that the aeration system worked effectively at all sites.

Faecal coliform counts were within expected ranges for primary treated effluent, except at the Woodpecker site were they were significantly lower. It is thought that a disinfection chemical either in the glass-washer or in a urinal cake is being used and thus lowering the faecal coliform count; however the exact mechanism has not yet been established. The Gem-Air and Pritchard Rd sites both had aboveground reuse of the treated effluent which involved a disinfection technology being added to the KEWT system; hence the low faecal coliform counts. The other four sites either had no discharge or conducted all recycling in a sub-surface manner that meant that the licence compliance was for <1000 cfu/100 ml and a specific disinfection technology was not required.

Representative data obtained from the SCADA technology are presented in Figure 2, showing the results of the hydraulic balancing.

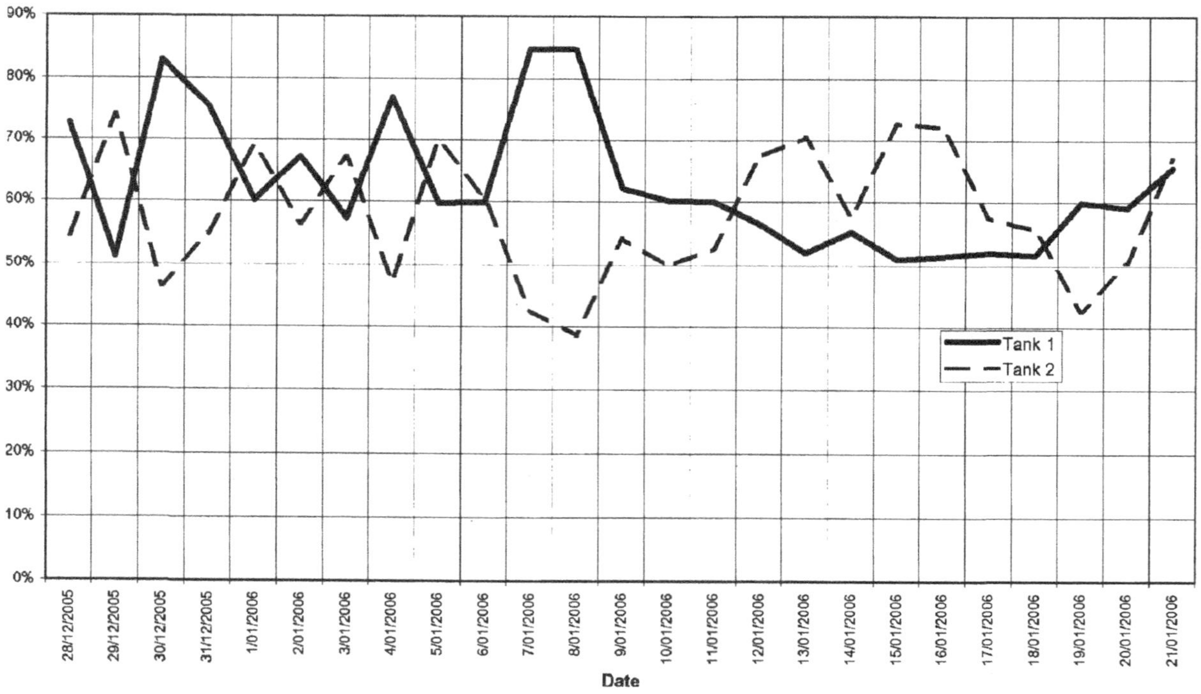

Figure 2.1.2: SCADA data from Burpengary Site for Tank 1 and Tank 2

Tank 1 represents the holding tank into which the primarily treated effluent flows into; tank 2 collects the non-transpired effluent and pumps it back into tank 1. The majority of the wastewater at the Burpengary site is produced between 7:00 am and 6:00 pm on weekdays. The site is not open at night or on weekends. Approximately 7000 L of wastewater (predominately blackwater) is produced per weekday. The only greywater produced at the site (estimated at less than 10% of total wastewater production) is through the activities of the cleaner, a small kitchenette and one shower. The 31/12/05, 1/1/06 and the 2/1/06 represented a Saturday, Sunday, and Public Holiday Monday; hence no flow and the reduction in tank volumes can be clearly seen. It rained over the weekend of the 7/1/06 and the 8/1/06 and as no evapotranspiration was occurring in the KEWT channels the SCADA system told the pumps to remain inactive and the tank volumes did not significantly change. While overcast on the weekend of the 14/1/06 and 15/1/06 evapotranspiration still did occur albeit at a reduced rate hence the small changes in tank volume. Over the 4 weeks represented in the figure approximately 140 000 L of wastewater was produced mostly within a 220 hour time period (out of a possible 672 hours). The SCADA system was able to balance the amount of effluent within each of the two tanks so that no overflows or hydraulic surges occurred at the Burpengary site, thus minimising the chances of the production of poorly treated effluent.

Alarms have been triggered, for example the unauthorised turning off of the main power switch and staff trying to flush disposable nappies which in turn caused pump blockages. The SCADA monitoring system sent alarm messages in both instances indicating a loss of power and a pump malfunction respectively. A maintenance worker was able to be immediately dispatched and the faults were repaired within three hours.

Discussion

The results in Table 2.1.2 showed that the primary wastewater produced at each of the sites was notably different due to site specific characteristics. The ratio of greywater to blackwater was different from site-to-site as well as being outside typical domestic ranges at some sites. While some wastewater characteristics such as DO showed very little variation from normal domestic values contaminants such as BOD_5 and TN did occur at relatively high concentrations. The EP volume may not be the ideal design criteria for commercial sites due to the wastewater characteristics and associated mass-loadings of contaminants of concern. Systems designed to treat a certain volume of wastewater may ineffectively treat a highly concentrated effluent with relatively large values of BOD_5 and/or TN, TP, etc. The mass-loadings values for different wastewater characteristics for specific activities, such as at a licenced premises, have not been well-researched and are not available from Australian regulatory authorities. To design on-site wastewater systems for commercial sites using mass-balance equations would currently be difficult due to the absence of definitive data. It may be worthwhile to research the most common commercial operations that use on-site wastewater treatment technologies and develop mass-balance guidelines based on the characteristics of the wastewater. The volume and the characteristics of the wastewater produced at commercial sites may require a stronger design and regulatory framework to ensure the treated effluent quality and the protection of public and environmental health.

The KEWT system was able to adequately treat the wastewater produced at all six sites. The combination of treatment technologies, such as the volcanic rock and sand filters, the addition of additional filter substrates such as charcoal, the microbial biofilms, the soil beds and the selected plants in the recirculating evapotranspiration channel all work in concert to reduce or transform the contaminants of concern. The recirculation of the non-transpired water allows for the additional aeration, detention and therefore treatment of the effluent, as well as offsetting possible adverse impacts due to hydraulic surges. The SCADA technology ensures that for the pumping cycles are adjusted to the real-time need of the system and allows for any operational problems to be immediately transmitted to a central location resulting in a prompt maintenance response. The KEWT system and its related support technologies can reduce the concerns associated with commercial on-site systems especially those relating to design, operation and maintenance parameters.

Conclusion

The designs and installations of commercial on-site wastewater treatment and reuse technologies should reflect both the volume of wastewater and the mass-loadings of the contaminants of concern to ensure that the treated effluent does not pose unnecessary risks to either public or environmental health.

Acknowledgements

The authors would like to recognise the efforts of the GBG team, Jeff, Paul, Dennis, Cameron, Jamie, Graham and Cathy, Donna Hobbs, Robert Lowry, Charmain Elder, Billy Sinclair, Nicola Orr, Linda Ahern of the PSG, and the Ecoscope team, and Pam McKennariey, and Gavin and Kathy Kele.

References

APHA (2001). *Standard Methods for the Examination of Water and Wastewater*. Washington, D.C., American Public Health Association.

Asquith, B., L. Shelly, *et al.* (2005). *Performance of Medium Scale On-site Wastewater Systems*. On-site 05, University of New England, Armidale, Lanfax Laboratories.

Fleeger, J. W., K. R. Carman, *et al.* (2003). "Indirect effects of contaminants in aquatic ecosystems." *The Science of The Total Environment* **317**(1-3): 207-233.

Kele, B. (2005). Onsite Wastewater Treatment and Reuse Using Recirculatory Evapotranspiration Channels In Regional Queensland. *PSG - Master of Applied Science*. Rockhampton, CQU: 1-338.

Kele, B., D. J. Midmore, *et al.* (2003). *On-site Wastewater Treatment And Effluent Reuse At An Ecotourism Resort*. Water Recycling Australia - 2nd National Conference, Brisbane, Australian Water Association.

McAvoy DC, D. A., Nielsen AM, Cano ML. (2002). "Investigation of an onsite wastewater treatment system in sandy soil: modeling the fate of surfactants." *Environ Toxicol Chem.* **21**(12): 2623-30.

Panswad, T. and L. Komolmethee (1997). "Effects of Hydraulic Shock Loads on Small On-site Sewage Treatment Unit." *Water, Science and Technology* **35**(8): 145-152.

Sostar-Turk, S., I. Petrinic, *et al.* (2005). "Laundry wastewater treatment using coagulation and membrane filtration." *Resources, Conservation and Recycling* **44**(2): 185-196.

Tchobanoglous, G. and F. L. Burton (1991). *Wastewater Engineering - Treatment, Disposal, Reuse*. Singapore, McGraw-Hill Book Co.

Tchobanoglous, G., F. L. Burton, *et al.* (2003). *Wastewater Engineering - Treatment, Disposal, Reuse*. Singapore, McGraw-Hill Book Co.

3 Bioreactors and Filters

3.1 Municipal Solid Waste Leachate Treatment and Retention by Cover Soil with Salt-Tolerant Grass

Wilai Chiemchaisri[1*], Chart Chiemchaisri[1], Warintorn Boonyaroch[1] and Sayan Tudsri[2]

[1] Department of Environmental Engineering, Faculty of Engineering, Kasetsart University, Thailand

[2] Department of Agronomy, Faculty of Agriculture, Kasetsart University, Thailand

(* Corresponding author: fengwlc@ku.ac.th)

Abstract

This study examined the application of tropical salt tolerance grass to the treatment of high organic strength municipal solid waste leachate. Two experiments were carried out to investigate the effect of wastewater with high organic and salinity on growth of tropical grass, treatment and retention of organic substances by soil with salt-tolerant grass. In the first experiment, pot study was used to find appropriate plant species which can survive under their exposure to high organic loading rate (OLR) of 0.05-0.5 kg COD/m^2.d. Subsequent experiment was conducted using lysimeter study to investigate leachate treatment efficiency of cover soil with salt-tolerant grass at various hydraulic loading rates of 1, 2.8, 5.6 cm/d.

In pot experiment, three different forage crops known as salt tolerant species were employed: *Pennisetum purpureum, Cynodon plectostachyrus, and Sporobolus virginicus*. The experimental results suggested that leachate irrigation stimulate plant growth if it is provided at appropriate application rate. *S. virginicus* could be grown at high OLR of 0.1 kg COD/m^2.d without significant damages where the others survived at maximum OLR of 0.05 kg COD/m^2.d. Application of higher OLR than 0.1 kgCOD/m^2.d caused damages to all species, mainly as a result of oxygen deficiency.

In soil lysimeter study, it was found that cover soil with *S. virginicus* yielded high efficiency of pollutants removal. Under the application of high strength leachate (8,500 mgCOD/L) at HLR 1 cm/d, more than 80% removal efficiencies of both BOD and COD could be achieved. In terms of nutrient removal, 63-76% and 48-50% removals of nitrogen and phosphorus were obtained respectively. Nevertheless, increase in HLR reduced treatment efficiency and plant growth rate. The results of this study showed that cover soil with *S. virginicus* could be successfully applied for the treatment of high strength leachate. It is a promising alternative for the treatment of leachate at small landfills in developing countries.

Key Words

Leachate, land treatment, salt-tolerant grass, landfill cover soil

Introduction

Landfilling is the common method for disposal of solid wastes when large land area is available. Under proper operation, landfilling of solid wastes could limit environmental problems such as odorous smell, infectious vectors, leachate pollution etc. Nevertheless, management of leachate remains one of the problematic operations anticipating in landfill. Usually, sophisticated system is needed to cope with the fluctuation of quantity and characteristics of this wastewater. More specifically, it also contains high content of hardly biodegradable organics, nutrients and salinity. In small landfills of developing countries where high mechanized treatment system is not affordable, recirculation of leachate into the landfill cell by irrigating it on cover soil during dry season can be one of low-cost solutions. While the waste layer can help eliminating some of the pollutants in leachate, cover soil also plays an important role in reducing its quantity and pollutant concentrations through natural evaporation and treatment capability in soil (Ettala, 1997). This operation may also adversely affect the plant grown on landfill cover soil (Maurice and Langerkvist, 1997). Irrigation of leachate can help maintaining soil moisture necessarily for vegetation but high salinity in leachate can also damage the plant by which it can be determined in terms of electrical conductivity (Soil and Plant Analysis Council, 1999). Provision of salt tolerant-plant on the landfill cover and optimum loading of leachate should be considered for the achievement of sustainable irrigation in landfill operation. This study was focused on the effects of leachate on different species of salt-tolerant grasses. The studies were conducted through pot and lysimeter experiments. Besides, soil property changes during leachate irrigation, i.e. soil EC and pH, and soil respiration, were also studied via another set of batch experiments.

Materials and Methods

Effect of leachate on plant growth

Optimum organic loading rate of leachate for plant was investigated in this study. Three species of salt-toterant grasses: *Pennisetum purpureum, Cynodon plectostachyrus, and Sporobolus virginicus* (Tudsri, 1997) were grown on sandy loam in 0.15x0.15 m sized pots. Each species was planted in triplicate at each organic loading rate. Initial height of plant was controlled at 10 cm. They were irrigated with rainwater every day during the first two weeks to obtain healthy condition before leachate application. Fresh municipal solid waste leachate (13,000-16,500 mgBOD/L, 16,000-19,000 mg COD /L, 400-800 mgTKN/L, 300-500 mgNH_3-N/L, 3-8 mgNO_3-N/L, 60-80 mgPO_4/L, pH 4.3-6.0 , EC 8 dS/m) diluted with tap water to obtain COD concentration of 1,700; 3,500; 8,800; and 17,700 mg/L and EC contents of 1.05, 2.06, 4.00, and 8.18 dS/m (10, 20, 50, 100% of leachate) were used. Each dilution was applied at a rate of 0.4 cm/d, therefore, the organic loading rates became 0.05, 0.10, 0.25, and 0.5 kgCOD/m^2.week. All tests had been conducted for 16 weeks. Plant growth was monitored by counting its leaves and height on weekly basis. Other visual plant damages such as yellow/brown leave, if any, were recorded. In control experiment, the plant was irrigated by rain water at the same application rate.

Effects of leachate on soil quality/soil respiration

Soil quality experiment: The accumulations of salt (EC) and organic matter (OM) contents in soils due to leachate irrigation were investigated in this experiment. Forty pots with 50-g of sandy clay loam soils each were prepared. Soil and leachate used in pot experiment were employed. About 2-mL of leachate at different concentrations was spiked into testing pot every day for eight weeks. This is equivalent to the same volumetric loading rates of leachate applied in the pot-experiment in which COD loading rates were varied at 0.05, 0.10, 0.25, and 0.5 kgCOD/m^2.week. Every week, soil samples from two pots were examined for their EC and OM contents.

Soil respiration experiment: This experiment was set up to determine the effects of leachate concentration on soil respiration as expressed in term of oxygen uptake rate (OUR). Ten grams of sandy clay loam soils were placed into 118-mL serum bottles. Various leachate concentrations were prepared by diluting raw leachate with distilled water. Leachate of 0.5-mL at each dilution was transferred into duplicate bottles to provide COD loading of 0.005, 0.007, 0.018, 0.036, 0.076, 0.134 and 0.2 kgCOD/m^2 and the bottles were closed with the aluminum-silicone caps. Oxygen content in headspace gas was then analyzed at different time interval by gas chromatograph (GC6890, Agilent; TCD, carrier gas: helium 65-mL/min; packing materials: CRT1, Alltech in stainless steel column) to determine OUR.

Treatment efficiencies of vegetated cover soils at different hydraulic loading rates

Three laboratory scale soil lysimeters (0.4x0.6x0.8 m^3) containing 0.60 m depth of sandy loam soil was conducted in order to find optimum hydraulic loading rate (HLR), and treatment efficiencies of vegetated cover soils. Two species of grasses, pre-determined from pot experiment, were selected, i.e. *S. virginicus and P. purpureum*. HLR of leachate (8,500 mgCOD/L) was set at 1, 2.8, and 5.6 cm/d (equivalent to organic loading rate of 0.08, 0.238 and 0.476 kgCOD/m^2.d). Leachate treatment efficiencies were determined in terms of COD, BOD, SS, NH_3-N, NO_3-N, TP and EC removals. Soil characteristics were also examined before and after each run.

Results and Discussions

Effect of leachate on plant growth

Figure 3.1.1 shows the variation of plant heights during leachate application at different loading rates (0-100%). The results show that plant growth was stimulated when irrigated with diluted leachate (10% leachate) during 15 weeks of experimental period. At higher leachate concentrations (20-100%), *P. purpureum* and *C. plectostachyus* were severely affected and died after 7th and 13th week respectively. Visual appearances of plants were yellow and brown leaves. Only *S. virginicus* could grow under 20% leachate with highest growth yield which indicating that it was the most tolerant grass in this study. Because the plant health condition will affect leachate treatment efficiencies, therefore it is recommended that diluted leachate should be used throughout the operational period, and organic loading rate should be controlled below the tolerable limit of each plant species.

Effect of leachate on soil quality and respiration rate

Because plant condition could be affected by soil pH, EC and OM content, the change of these parameters in soil was monitored (Figures 3.1.2, and 3.1.3). From the experimental results, it can be seen that diluted leachate (0-50%) had little effects on soil EC, whereas concentrated leachate increased soil EC significantly after six weeks of leachate spiking. The determination of soil pH showed that it was lowered when applied leachate quantity was increased. Soil pH turning acidic especially when irrigated with concentrated leachate (100%).

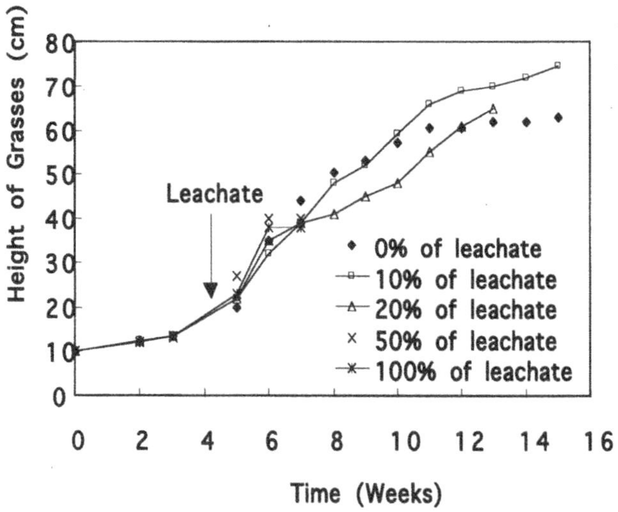

(a) *P. purpureum*

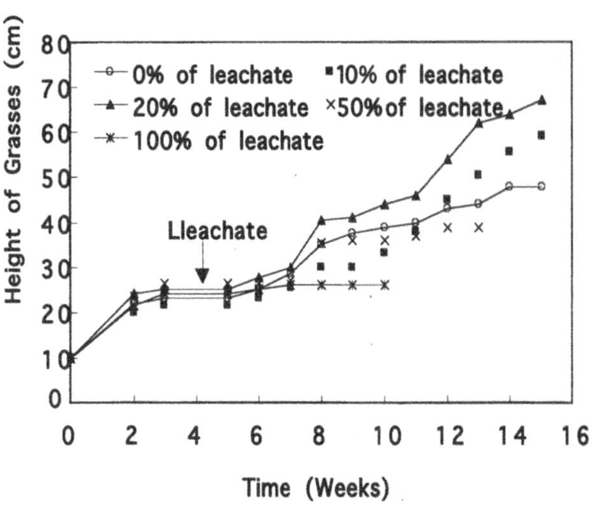

(b) *S.virginicus*

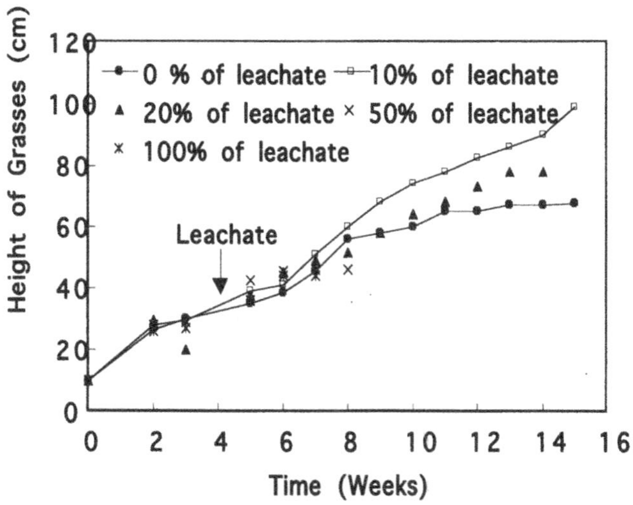

(c) *C. plectostachyus*

Figure 3.1.1: Growth of various plant species in different loading of leachate (a,b,c) (note: discontinued plots were used after plant death).

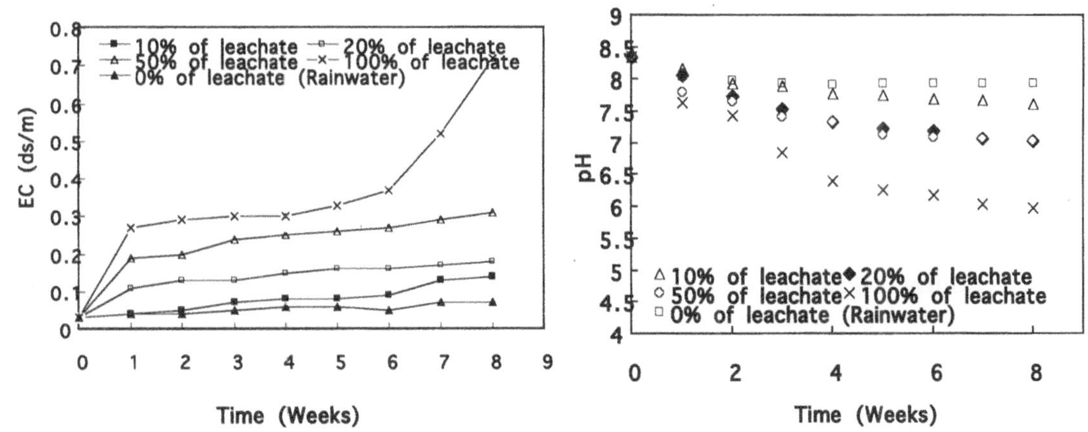

(a) EC (b) pH

Figure 3.1.2: Changes of EC (a), pH (b) at different loadings of leachate in the soils

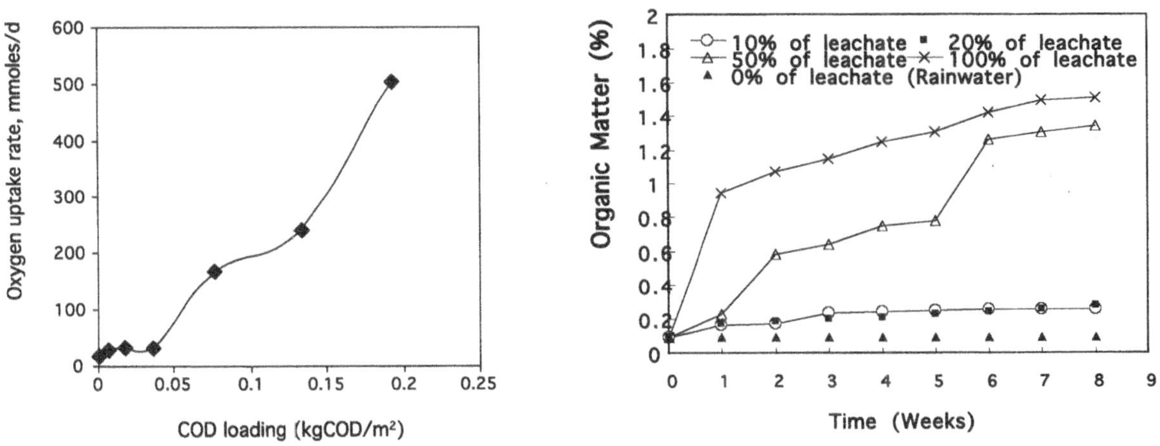

Figure 3.1.3: Changes of oxygen uptake rates (a) and organic matter (b) at different loadings of leachate in the soils

In case of oxygen uptake rate (OUR) as shown in Figure 3.1.3, the result indicates that OUR in soils increased with an increase in COD loading rates. Rapid increase in OUR was observed when COD loading rate was higher than 0.036 kg COD/m^2. A maximum OUR of 510 mmolO_2/d was detected in soil at COD loading of 0.2 kgCOD/m^2. Oxygen deficient condition could be occurred in soil when irrigated with high strength leachate (0.25-0.5 kgCOD/m^2).

High contents of EC, COD and nutrients in leachate could be the possible reasons of plant damage when vegetated cover soil was irrigated with concentrated leachate. Whereas pH was unlikely to be a key factor as all plants could be grown with acidic leachate (pH of 4.29-6.0). In general, level of soluble salts in soil can provide significant impact on plant growth due to nutrient imbalance. The growth of sensitive plants is normally restricted in soil with EC in range of 0.8-1.2 dS/m as compared to 1.6-3.2 dS/m of some tolerant plants (Soil and Plant Analysis Council, 1999). In this study, death of grasses appeared even under salt accumulation range (1.8-2.4 dS/m) below reported level for plant damage. Therefore, soil salinity should not be the only factor affecting plant death in our pot experiment. Other possible factor should be the accumulation of organic matters which subsequently causing rapid increase in OUR and oxygen deficient condition in soils.

Root respiration was affected by oxygen availability in soil. The oxygen demand of plant root was reported in range of up to 1,600 ml O_2/kg fresh roots/h (Glinski and Stepniewski, 1986). This oxygen limited condition which causing plant death is known as asphyxia (Gendebien *et al.*, 1992). Maurice and Langerkvist (1997) reported that oxygen content down to 5-10% in soil had negative effect on vegetation in landfill. Asphyxia of vegetation is also affected by deficiency in soil minerals e.g. K, N, P, Ca and Mg (Flower *et al.*, 1981; Leone *et al.*, 1997). Thus, leachate irrigation at appropriate organic loading rate not only prevents oxygen-deficient condition but also provides additional nutrients for plant growth.

Treatment efficiencies of vegetated cover soils at different hydraulic loading rates

Treatment efficiencies of bare and vegetated soil and soil characteristics before and after leachate application were shown in Table 3.1.1 and 3.1.2, respectively. The results showed that, at HLR of 1.0 cm/d and leachate concentration of 8500 mgCOD/L, more than 80% removal efficiencies of BOD and COD was obtained in vegetative soil lysimeters with *S. virginicus* and *P. purpureum*. Average organic removal in bare soil lysimeter was 72%. When HLR was increased to 2.8 cm/d, COD and BOD removals efficiencies in both vegetated soil lysimeters slightly reduced to 75-78 % and 79-81%. Only *S. virginicus* survived when HLR was further increase to 5.6 cm/d, however, removal efficiencies were lowered to 58% for BOD and 68% for COD. In case of *P. purpureum,* the plant died after two weeks of leachate irrigation at this highest application rate (5.6 cm/d). Their treatment efficiencies were then reduced to that comparable with bare soil (35%BOD and 53%COD removal). These results suggested that vegetation plays significant role in organic removal mechanism in the system. In terms of nutrient removals, nitrogen and phosphorus removals were 63-76% and 48-50% in vegetated soil lysimeters at HLR of 1 cm/d. Similar to organic removal, the removal efficiencies decreased when HLR increased. It can be concluded that vegetative soil provided better treatment efficiencies as compared to bare soil. *S. virginicus* was found to be best for landfill vegetation in this study with the optimum HLR of 1.0 cm/d.

Table 3.1.2 shows the accumulation of salts, organic matters and total nitrogen in soil at HLR of 1.0 and 2.8 cm/d. It appears that higher accumulation of those substances was found in the bare soil than that in the vegetated soil lysimeters. Nevetheless, slight differences among two vegetated lysimeters examined in this study was observed. It is likely that microbial degradation of organic matter in rhizosphere were responsible for this higher organic removal efficiency. About 2-3 times of salt removal was also observed in vegetative soils as compared to bare soil as the plants could uptake nutrients and other minerals for their growth. The results also suggest that soil salinity was not the major factor causing plant growth in this study.

From this experiment, vegetated soil with salt-tolerant grasses was found effective in removal of organic and nutrients from leachate if it is applied at appropriate loading rate (1 cm/d in this study). It is considered a promising leachate management options for small landfills in developing countries.

Conclusion

In pot experiment, leachate irrigation stimulated forage crop growth if it is provided at optimum application rate, e.g. organic loading rate (OLR) of 0.1 kg COD/m^2.d for *Spoloborus virginicus,* without significant damages. For *Pennisetum purpureum, Cynodon plectostachyrus,* a maximum OLR of 0.05 kg COD/m^2.d should be controlled to avoid oxygen deficient condition. Organic matter but not salinity was an important factor affecting plant survival. In soil lysimeter study, only *S. virginicus* could grow well at hydraulic loading rate (HLR) of 1.0-5.6 cm/d. More than 80 % removal efficiencies of BOD and COD could be achieved at HLR of 1 cm/d. In terms of nutrient removals, 63-76% and 48-50% removals of nitrogen and phosphorus were achieved by vegetated soil. Nevertheless, increase in HLR reduced treatment efficiencies and plant growth yield. This study suggests that cover soil with *S. virgincus* could be applied to the treatment of high strength leachate. It is a promising leachate management options for small landfills in developing countries.

Table 3.1.1: Treatment efficiencies of vegetated soil with S. virginicus (SV) and P. purpureum (PP) and bare soil (control) at different HLR

Parameters	Influent	HLR 1.0 cm/d			HLR 2.8 cm/d			HLR 5.6 cm/d		
		Eff- SV	Eff- PP	Eff -Control	Eff-SV	Eff-PP	Eff -Control	Eff-SV	Eff-PP	Eff -Control
SCOD, mg/L	8460±103	1494±25	992±6	2382±5	1828±3.6	2102±11	3407±7.5	3597±69	5426±33	5438±47
SBOD, mg/L	6177±185	1247±32	754±4	1777±5	1166±4	1275±5	2020±38	1946±46	2899±12	2906±11
TKN, mg/L	244.5±6.5	89±4	84±0.3	104±0.4	131±4	143±7	184±10	156±8.5	207±3	212±3
NH_3-N, mg/L	99.9±1.4	24±0.8	36±4.2	34±0.5	35±2.7	37±0.9	76±3.6	57±0.5	78±0.6	80±0.6
NO_3-N,mg/L	1.8±0.03	0.56±0.03	0.81±0.02	0.74±0.01	1.0±0.1	1.1±0.1	1.4±0.1	1.25±0.05	1.45±0.04	1.46±0.04
NO_2-N, mg/L	1.7±0.03	0.36±0.03	0.7±0.04	0.77±0.02	0.6±0.1	0.7±0.1	0.95±0.1	1.0±0.06	1.22±0.05	1.24±0.05
TP, mg/L	28±1.0	14.0±0.6	14.6±0.5	14.8±1.0	16.7±1.0	17.3±0.3	19.8±0.2	17.8±0.35	21.5±0.2	21.9±0.25
SS, mg/L	525±4.62	108±10	24±3.5	25±0.7	255±5.6	263±8.3	331±6.6	373±4.5	409±6.0	418±3.5
pH	4.29±0.5	7.59±0.07	7.26±0.06	7.89±0.02	7.75±0.05	7.8±0.04	8.2±0.04	7.8±0.12	8.3±0.08	8.36±0.06
EC, dS/m	7.74±0.05	10.50±0.03	12.35±0.06	15.18±0.04	12.22±0.9	12.36±0.9	15.37±0.76	12.68±0.9	16.47±0.8	16.71±0.9

Table 3.1.2: Comparison of soil characteristics before and after leachate irrigation

Parameters	HLR 1 cm/d				HLR 2.8 cm/d			
	Before	After- SV	After- PP	After -Control	Before	After-SV	After-PP	After -Control
pH	8.1	7.6	6.8	7.2	8.2	7.6	6.8	7.2
EC, dS/m	0.03	5.69	5.81	13.72	0.19	14.53	15.67	21.25
OM,%	0.095	0.90	1.04	1.60	0.15	1.55	1.59	1.95
TN,%	0.02	0.15	0.16	0.20	0.04	0.23	0.25	0.29

Acknowledgement

This study is conducted in Asian Regional Research Program on Environmental Technology (ARRPET) supported by Swedish International Cooperation Development Agency (SIDA).

References

Ettala M., (1987). Influence of Irrigation with Leachate on Biomass Production and Evapotranspiration on a Sanitary Landfill, *Aqua Fennica*, 17(1), 69-86.

Flower, F.B., Gilman, E.F., and Leone I.A., (1981). Landfill Gas. What it does to trees and how its injurious effect may be prevented?, *Journal of Arboriculture*, 7, 43-52.

Gendebien, A., Pauwels, M., Ledrut-Damanet, M.J., Nyns, E.J., Willumsen, H.C., Bustson, J., Fabry R., and Ferrero, G.L., (1992). Potential Landfill Gas Damages to Vegetation. In: Landfill gas from environment to energy, Luxembourg, Commission of the European Communities, 35-36.

Glinski, J., Stepniewski, W., (1986). Soil Aeration and Its Role for Plants, CRC Press Inc. Boca Raton, Florida, USA

Leone, I.A., Flower, F.B., Arthur, J.A., and Gilman, E.F., (1997). Damage to Woody Species by Anaerobic Landfill Gas, *Journal of Arboriculture*, 3, 221-225.

Maurice, C and Langerkvist, A., (1997). Landfill Gas Properties and Effect on Green Plants-Literature Review, Division of Landfill Science & Technology, Lulea University of Technology, Sweden.

Soil and Plant Analysis Council Inc., (1999). Soil Analysis Hand Book of Reference Method, CRC Press, Washington D.C., p 60.

Tudsri, S., (1997). Tropical Forage Crops: Production and Management, Kasetsart University, Bangkok, Thailand. (in Thai).

4 Systems for Developing Communities

4.1 Development of an Aerobic Vermiculture Whole-of-Waste Treatment Plant and potential uses in Sub-Saharan Africa

S.J.Riley, A.Panikkar, S.Shrestha
School of Engineering
University of Western Sydney
Locked Bag 1797, Penrith South DC, NSW 1797, Australia

Abstract

The management of municipal and household waste in Sub-Saharan Africa is a difficult problem as the government and individuals try to manage the costs associated with systems expected to be the norm in developed economies with increasing demands for lifestyles typical of the west. In Kenya and Uganda the management of municipal and human waste is basic, if present at all, and usually concentrated in the largest cities.

Increasing population densities in many areas mean that the traditional pit latrine and septic systems are no longer environmentally acceptable, as groundwater contamination and overflows during periods of heavy use and in the wet seasons contaminate the existing water sources of springs, wells and streams.

The need for a simple and low cost waste management system that incorporates all the organic waste of households and small communities is an urgent necessity. At the University of Western Sydney a prototype of a system based on vermiculture was trialled and shown to be effective in generating products that could be used by the community: worms for feedstock to poultry and fish, compost for gardens, and water suitable for irrigation.

This paper presents some of the design characteristics of the system. Future development work is outlined.

Introduction

There is a well-recognised need to manage the waste of communities in developing areas in order to reduce disease and improve the quality of life (Sato, 2004). Communities should be able to value add from the management of waste, by utilising by-products from the processing of waste.

At the University of Western Sydney the Sustainable Engineering and Technology Research Group has been working on vermicomposting of the putrescible waste of families and small communities. We argue that vermicomposting has the potential to deliver water that meets World Health Organisation standards for irrigation of crops, worms that can be harvested and used for aquaculture or poultry feed, and compost that is a rich fertiliser.

Over the course of three years a prototype was developed and trialled in laboratory situations. A number of papers report on the results of these trials (Okalebo and Riley, 2003; Panikkar and Riley, 2003; Riley, Panikkar and Okalebo; 2003; Panikkar, Riley and

Shrestha, 2004; Panikkar *et al.*, 2005; Panikkar, Hackney and Riley, 2005) and the system is briefly described in the following.

The main purpose of this paper is to outline some of the issues that we believe have to be addressed in introducing such technology into sub-Saharan Africa.

The Need

The traditional pit latrine and the more expensive septic tank are rapidly becoming less desirable options for management of human waste in certain parts of Africa. Contamination of waterways and wells as a result of poorly located treatment systems, and the increasing population of some areas stressing the existing systems beyond their capacity are well known. Movement of populations into less desirable areas, such as swamps, reduces the viability of simple treatment systems. Yet there is little choice for many communities because the more expensive treatment systems are not affordable, or there are insufficient management resources or infrastructure that would enable such systems to work effectively. The disposal of human waste is just part of the waste cycle in all communities. Putrescible material, such as food waste, paper and vegetable matter of one sort or another has to be disposed of. The piles of such material, often intermingled with other wastes, is a common site in many slum areas of the developing world.

The other scene often presented in poorer areas is wastewater, flowing through the rotting rubbish and along the streets (Fig 1). This water often has its source in either overflowing latrines or is grey water disposed of in the streets outside residences.

Figure 4.1.1: Wastewater in the streets of a poor district of Kampala and solid waste in India

Waste is recognised as a resource and so the opportunities to use this waste should be realised. A waste that becomes a resource can only add to the quality of life of the community that grasps the opportunity to use the resource (Bernache, 2003; Buenrostro and Bocco, 2003; Ambulkar and Shekdar, 2004).

At UWS we considered that it was possible to turn the putrescible waste and all household water, including urine, into a resource that the community could use. Vermicomposting is a logical option as it is a well-known technique in agricultural communities. Thus there is the initial advantage that communities have some skill in the area and are familiar with the processes of composting and the role that worms may play in this. Worms can be found in almost all environments, and thus are a readily obtainable resource. There are well-known uses of worms in agriculture and the value of compost and worm castings as a fertiliser are also well known (Lionello and Fransesco, 1989; Diaz *et al.*, 1993).

Microbial Composting and Pathogen Removal

An important aspect of biological waste treatment is the reduction of pathogens such as disease-causing bacteria, viruses, protozoa etc. In microbial composting, pathogen destruction is a result of competition with composting microbes for the available nutrients, high temperatures encountered due to thermophilic stages and some of the intermediate metabolites secreted by microbial respiration (Farrell 1993; WaterWatch 1999; Sidhu, Gibbs *et al.* 2001).

Certain species of earthworms (epigeic – litter dwelling) grow and consume organic waste more rapidly compared to other anecic (sub-surface) and endogeic (deep soil) earthworms (Aranda, Barois *et al.* 1999). Compared to microbial composting, vermicomposting is faster as worms and microbes both take part in digesting the waste materials (Edwards 1988; DNR 2001).

Previous studies have shown that vermicomposting can achieve safe levels of pathogen removal, by the intestinal enzymes and actions of worms and the beneficial soil microbes that the worms leave behind, with which the pathogenic organisms have to compete for the limited resources (Dominguez *et al.* 1997; Eastman 1999; DNR 2001). The hazards caused by pathogenic content in the waste materials and composting itself, and the risks thereof have been the subject of many discussions on public health, environmental safety and regulations.

Worms have been known to secrete certain fluids with antibacterial properties (Satchell 1983). Most human pathogens, including E. coli are killed by composting worms such as the most-widely used tiger worm (*Eisenia foetida*) and red worm (*Lumbricus terrestris*). Competition between pathogenic organisms and indigenous microflora in the worm castings for nutrients has been shown to cause pathogen removal (Rouelle 1983; Allievi, Citterio *et al.* 1986; Morgan 1988; Eastman 1999; Sidhu *et al.* 2001).

The Vermicomposting System

Following detailed tests on vermicomposting to validate results found in literature regarding the worm-stocking rate and feeding capacity, a prototype vermicomposting unit was designed and tested at the Sustainable Engineering and Technology (SET) research facility at the University of Western Sydney. The test results are presented elsewhere (Panikkar, Shrestha *et al.* 2003; Panikkar and Riley 2003; Panikkar and Riley 2003). The effluent from the prototype treatment plant was initially expected to provide three-order of magnitude reduction in pathogen counts, against comparable secondary treatment in centralized wastewater management of one-order reduction (WHO 2003).

The prototype incorporated three chambers for vermicomposting, all with solid materials at different stages of decomposition. The conceptual design of the treatment system is provided in Riley *et al* (2003). The prototype was tested by feeding a fixed amount of pig manure and organic material each day into the collection chamber. The quantity and composition of material fed into the reactor was based on an assumption of the quantity and quality of waste produced by a household, but scaled down for the size of the reactor. Details of the experimental design are given by Panikkar *et al.* (2003). The liquid waste (pig manure initially), which was used as the most closely comparable substitute for sewage, flows through the different chambers (Fig 2) with a designed hydraulic retention time (HRT) of maximum 1 day based on calculations of waste material quantities. The final effluent is collected at the end of a composting cycle and tested for faecal coliform, *E. coli,* several chemical and physical water quality parameters including turbidity, ammonium, nitrate and solids content. For quality control, occasional paired, spiked and blind samples were tested at a NATA accredited professional water-testing laboratory. Microbiological analysis was undertaken at the Australian Government Analytical Laboratory. The results of the prototype test were used in assessing the risk levels of associated events identified as of particular significance.

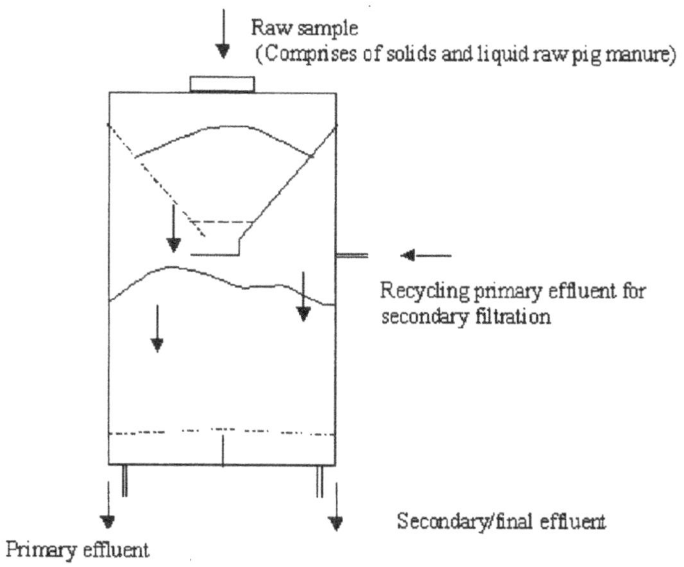

Figure 4.1.2: A schematic of the prototype with wastewater flow directions.

Risk Analysis

The risk evaluation matrix was applied to a selected set of events arising from possible hazards, with reference to the effects of pathogen content and public health. Most hazards could be avoided by proper maintenance of the system and basic personal safety such as use of personal protective equipment (gloves, dust mask and proper footwear). The result of the risk analysis of the vermicomposting system was compared against risk analysis of the direct blackwater application in agriculture and aquaculture. As expected, the direct application of nightsoil/blackwater to crops and fishponds poses a high risk of disease in most of the identified events compared to medium to acceptable/low levels for the vermicompost (Panikkar *et al.,* 2004).

Processes

The type of food material added and its moisture content did have an effect on the progressing of the worm-action in terms of the time taken for the worms to start consuming it. Worms usually devour a number of microbes that degrade the waste materials along with semi-degraded materials. The degradation of the material is then completed with the enzymes produced by the worms and the microbes promoted by these enzymes. To start this process, microbial action needs to have started on the material, which will depend on the moisture content of the material and the structure and particle size of the material. It was noted that pieces of sturdy materials such as wood remained in the substrate without much change, whereas other more putrescible material such as banana peels and cooked/uncooked vegetable parts degraded more quickly. Some materials such as well-moistened coconut fibre and smaller items such as rice were consumed faster while large pieces of bread and pasta were given lower preference. This could be due to the nature of bread that allowed growth of fungi (drier conditions – some are toxic to worms) and acidic pH of pasta. Contrary to many published and *ad hoc* reports on vermicomposting, the mixed breed of worms were seen not to be averse to meat and fish wastes that were supplied in semi-cooked and cooked conditions.

Degradation of paper shreds and garden organics was effected by the presence of other more humid waste materials such as kitchen waste and manure. Towards the end of the process, only tiny pieces of the long paper shreds remained and the dry leaves had all been converted into vermicastings. Small sticks and wooden pieces still remained, which had higher lignin content and thus were too strong for the microbial community and worms to act on them.

Addition of cow manure and duck droppings contributed to the performance of the worms, perhaps due to the higher organic nitrogen content and digested/degraded materials usually contained therein. Addition of materials richer in nitrogen was the preferred method of improving the C: N ratio.

Waste materials were added to the substrate in different patterns, such as in long strips or at corners or at different spots. This was done to check the response of the worms to different food materials (different pH, composition and structure). In general, the worms were seen to migrate away from the new material at first, only to return later and start the consumption. This might be due to the fact that worms cannot chew or consume any waste materials directly; or they need microbial action to start degrading the material. Once microbial action appeared the worms almost always returned to the newly added material. This was observed for all the materials except when coconut fibre was added at first on the bedding material. The worms appeared to be immediately attracted to the coconut fibre.

If the freshly added material lacked adequate moisture content fungi started to grow, which repelled the worms. This was reversed once water was sprayed over the material or moist materials were added. Certain fungi could be toxic to the composting worms, but other fungal predators such as nematodes would flourish once the numbers of fungal species increase and these are also part of the composting cycle.

The different worm species have different preferences and behavioural patterns, and this could be of use in successful operation of larger vermicomposting systems. Red wigglers were seen to grow in size more than all other species though red tigers were more prolific.

Benefits of the Vermicomposting Treatment

A safe waste treatment system such as vermicomposting of blackwater and biodegradable waste could be of use in places where centralised or other waste management facilities are unavailable. In vermicomposting, the waste materials are safely converted into soil amendment material with reduced levels of pathogens. Harvesting of worms as a protein source in fish culture has been in practice in many places, and this could be an application of vermicomposting in aquaculture. The liquid effluent from the vermicomposting unit could be further treated and utilised in irrigation or elsewhere. Proper treatment of effluent along with greywater from the household would offer a source of recycled water for different uses at the household, without adverse health implications.

Application in Sub-Saharan Africa

The successful completion of laboratory experiments at UWS has opened up the opportunity to conduct trials in Sub-Saharan Africa, one of the research areas for the Sustainable Engineering and Technology Research Group. The issues that need to be addressed are:

1. identification of suitable local worm species. This is likely to be a regional issue because of the considerable variation in climate and soils across the region. So it is not expected that one-size will fit all locations.
2. optimum building materials. The prototype was built from stainless steel in a well equipped workshop with highly experienced technicians. There is a major project in identifying alternative materials. Use of mud-bricks, with cement as a tackifier, and plastics is a possibility. Local expertise in timber and metal work will have to set the parameters for what is possible.
3. community expectations and management training. No system is completely free of maintenance, and there will be a need to ensure that the systems are properly maintained to give maximum benefits. There will also be a need to address community and cultural issues about the use and management of these systems. This is not expected to be easy and demonstration projects will have to be used.

We understand that there have been attempts to use vermicomposting systems in Sub-Saharan Africa, but at the time we undertook this study could not find any major reports on experiments and application. The opportunities are significant in both rural and urban communities, and there certainly is no operational knowledge of such systems amongst the people we have talked with in Uganda and Kenya.

One of the great advantages of a vermicomposting system that was widely adopted is that it would almost immediately reduce the quantity of garbage in many areas, thus relieving local governments of some of the burden of garbage collection and improving the visual amenity of communities.

Conclusion

Vermicompositing is a well established technique for managing the whole of waste in local communities. It has many spin off benefits that enable waste to be considered as a resource, and thus will probably sustain itself. It is relatively low cost to maintain. The issue we face is designing a system that is both easily and cheaply produced in developing countries and at the same time is acceptable to the communities that could use it.

References

Allievi, L., B. Citterio and A. Ferrari 1986. Vermicomposting of rabbit manure: modifications of microflora. *In*: Compost: Production, Quality and Use: Proceedings of a symposium organised by the Commission of the European Communities, Directorate-General Science, Research and Development., Udine, Italy, publ: Elsevier Applied Science.pp.115-126.

Ambulkar, A.R. and Shekdar, A.V. 2004. Prospects of Biomethanation technology in Indian context: a pragmatic approach. Resources, Conservation and Recylcing, 40(2), 111-128.

Aranda, E., I. Barois, P. Arellano, S. Irissón, T. Salazar, J. Rodríguez and J. C. Patrón 1999. "Vermicomposting in the tropics". *In: Earthworm management in tropical agroecosystems.* (Ed) P. Lavelle, Brussaard, L. & Hendrix, P. CAB International Publishing, Oxon, UK.

Bertmier, H. C. 2003. Garbage, Work and Society. Resources, Conservation and Recylcing, 39,193-210

Buenrostrop, O and Bocco, G. 2003. Solid waste management in municipalities in Mexico: goals and perspectives. Resources, Conservation and Recylcing, 39, 251-263.

Darwin, C. 1945. *The formation of vegetable mould - through the action of worms with observation on their habits.* Faber and Faber Limited, London.

Diaz, L. F., Savage,G. M., Eggerth, L. L., and Golueke, C. G. 1993. Composting and Recyling of Municipal Solid Waste. Lewis Publishers, Florida, USA.

DNR 2001. "Worm composting system". Department of Natural Resources, State of Missouri, USA.: Internet webpage *www.dnr.state.mo.us.* Accessed: 24-7-01.

Dominguez, J., C. A. Edwards and S. Subler 1997. "a comparison of vermicomposting and composting." worm world inc.: Internet webpage *www.gnv.fdt.net/~windle/refrence/index.htm.* Accessed: 26-7-01.

Eastman, B. 1999. Achieving pathogen stabilisation using vermicomposting. Biocycle(Nov 1999): 62-64.

Edwards, C. A. 1988. "Breakdown of animal, vegetable and industrial organic wastes by earthworms". *In: Earthworms in waste and environmental management.* (Ed) C. Edwards, & Neuhauser, F. SPB Academic Publishing, The Hague: 21-31.

Farrell, J. B. 1993. Fecal pathogen control during composting. *In*: International composting research symposium: Science and Engineering of Composting: Design, Environmental, Microbiological and Utilisation Aspects., Ohio, USA., publ: Ohio State Universitypp.282-300.

Lionello, B. and D. Z. Fransesco 1989. Soil fertility improvement and pollution risks from the use of composts referred to N, P, K & C balance. *In: Acta Horticulturae: International Symposium on compost recycling of wastes.* (Ed) C. Balis. Athens, Greece, International Society for Horticultural Science. 302.

Morgan, M. H. 1988. "The role of microorganisms in the nutrition of Eisenia foetida." *In: Earthworms in waste and environmental management.* (Ed) C. Edwards, & Neuhauser, F. SPB Academic Publishing, The Hague: 71-82.

Okalebo, S and Riley, S. 2003. Development and trial of a low-cost aerobic greywater treatment system. *In* K. S. Makhanu and C. C. Kamau (eds). Proceedings of the International Civil Engineering Conference on Sustainable Development in the 21st Century, Nairobi, Kenya, 12-16 August 2003. Institution of Engineers Kenya. p.328-335

Panikkar, A. and Riley, S. 2003. Issues in blackwater reuse – an example of a treatment System. *In* K. S. Makhanu and C. C. Kamau (eds). Proceedings of the International Civil Engineering Conference on Sustainable Development in the 21st Century, Nairobi, Kenya, 12-16 August 2003. Institution of Engineers Kenya. p.336-340.

Panikkar, A.K., Riley, S. J., Shrestha, S. P. 2004. Risk management in vermicomposting of domestic organic waste. Environmental health, 4(2),11-19.

Panikkar, A. K., Okalebo, S. A., Hackney, P. A., Shrestha, S. P., Riley, S. J. 2005. Household whole-of-waste management. Water 32(7),54-59.

Panikkar, A. K., Hackney, P. A., Riley, S. J. 2005. Degradation of solids during filtration of blackwater in a prototype vermicomposting unit. The Environmental Engineer, 6(1), 10-14.

Riley, S., Okalebo, S., and Panikkar, A. 2003. Multi-dimensional management of wastewater. *In* K.S. Makhanu and C.C. Kamau (eds). Proceedings of the International Civil Engineering Conference on Sustainable Development in the 21st Century, Nairobi, Kenya, 12-16 August 2003. Institution of Engineers Kenya. p.322-327

Rouelle, J. 1983. "Introduction of amoebae and *Rhizobium japonicum* into the gut of *Eisenia foetida* (Sav.) and *Lumbricus terrestris* L." *In: Earthworm Ecology: from Darwin to vermiculture.* (Ed) J. E. Satchell. Chapman and Hall Ltd, Cambridge, UK: 375-382.

Satchell, J. E. 1983. "Earthworm microbiology". *In: Earthworm Ecology: from Darwin to vermiculture.* (Ed) J. E. Satchell. Chapman and Hall Ltd, Cambridge, UK: 351-364.

Sato,S. 2004. Human security. 292 Lecture Series, UN Association.

Sidhu, J., R. A. Gibbs, G. E. Ho and I. Unkovich 2001. The role of indegenous microorganisms in suppression of *Salmonella* regrowth in composted biosolids. *Water Research.* **35**(4): 913-920.

WaterWatch 1999. "Human Enteric Pathogens in Sewage Sludge and Associated Disease". WaterWatch: Internet webpage *http://members.aol.com/wwanglia/frame_pathogen.htm.* Accessed: 30/08/2002.

5 Wastewater Quality

5.1 Design Considerations of Floating Plastic Media Pre-Filter for Removal of Particles from Surface Water with High Turbidity Level

Chart Chiemchaisri and Thanyarat Chomsuan
Department of Environmental Engineering, Faculty of Engineering,
Kasetsart University, 50 Phaholyothin Road, Chatuchak, Bangkok 10900, Thailand

Abstract

This study was conducted to determine design criteria of floating plastic media bed used for pretreatment of surface water with high turbidity level. It was performed by using synthetic raw water with constant turbidity of 80 NTU. From the jar test, polyaluminum chloride was found to be the best coagulant with an optimum dose of 15 mg/l. Filtration test using floating plastic media bed with different media size of 3,6,8 mm and bed depth of 0.4, 0.6, 0.8 m at filtration rate of 5, 7.5 and 10 m^3/m^2.h was examined. Highest particle removal efficiency was achieved when the system was operated using 6 mm media size and 0.6 m bed depth at 5 m^3/m^2.h filtration rate. The size of particles released from the filter increased with time. Proposed criteria to achieve 80% turbidity removal in this floating plastic media pre-filter were L/d of 100 or more, bed depth of 0.6 m or more with filtration rate between 5 and $10m^3/m^2$.h. The system was then applied at a water treatment plant in Bangkok where raw water turbidity fluctuated between 75-166 NTU. Average turbidity in treated water from floating plastic media filter was 7.5 and 28.1 NTU at the filtration rates of 5 and $10m^3/m^2$.h.

Introduction

Floating plastic media bed is applied as a simple water treatment system to remove suspended particles by contact-flocculation filtration mechanisms (Ben Aim *et.al.*, 1993; Chiemchaisri *et.al*, 2003; Ngo and Vigneswaran, 1995; Visvanathan *et.al.*, 1996). In small community water supplies, the floating medium filter can be used in place of conventional flocculation and sedimentation. It operates in a down-flow mode, under constant head pressure and incorporates in-line flocculation. Coagulated particles enter the filter where inter-particle contacts are promoted as the water flow through floating media bed. Flocculated particles are trapped in the bed where the separation takes place. The system could be operated at high solid retention capacity and low headloss development. Our previous investigation has shown that the floating plastic media bed of 1.50 m. depth has high turbidity and suspended solids removal efficiencies of more than 90%, producing low turbidity effluent of less than 5 NTU while having headloss of only 0.35 m of water after continuous operation for 24 hours (Chiemchaisri *et.al.*, 2003).
The floating plastic media bed was also found to be an effective pre-treatment unit for microfiltration membrane in an integrated water treatment system by increasing the size of particles entering the microfiltration unit to some extent and thus reducing the clogging of microfiltration membrane (Chiemchaisri *et al.*, 2005).

Despite high removal efficiency in floating plastic media bed, the design of media size and depth need to be optimized in order to obtain high degree of flocculation and filtration in the system. In this study, the floating plastic media size, bed depth and filtration rate were varied in order to obtain the best combination and optimum operating condition for particle enlargement through agglomeration and removal by filtration mechanism. Design criteria for the design and operation of floating plastic media bed were then proposed.

Materials and Methods

Schematic diagram of experimental system is shown in ***Figure 5.1.1***. Raw water was supplied into the filter column through a head tank. After coagulant has been added, coagulated water was fed into an acrylic column of 9 cm. in diameter and 160 cm. in length. Polypropylene beads of 3, 6 and 8 mm. size (P = 0.87 g/cm^3) were used. The depth of floating plastic media and coarse sand was varied at 40, 60 and 80 cm. The headloss in the system were monitored by a manometer.

Two types of coagulant were used in laboratory scale experiment, i.e. alum and polyaluminum chloride (PACl). They were prepared as stock solution at 1% concentration before being fed into raw water by a chemical dosing pump. Synthetic raw water made from kaolin clay having turbidity of 80 NTU was used in the first experiment. Subsequent experiment was then conducted at Bangkhen water treatment plant using actual raw water of Bangkok Metropolitan Water Works Authority. The chemicals used in actual operation of the water treatment plant , i.e. alum, ferric chloride, PACl, lime and polymer, were tested in the system.

The system was operated at different media sizes and filtration rates. The optimum dose of each coagulant was determined by standard jar test procedure. Effect of chemical dose (20%, 50% and 100% of jar test) was also examined in the experiment with actual surface water. The filtration rate of system was tested at 5, 7.5 and 10 m^3/m^2.h. Particle size in treated water was also studied using microscopic observation method (Olympus microscope, CHS model).

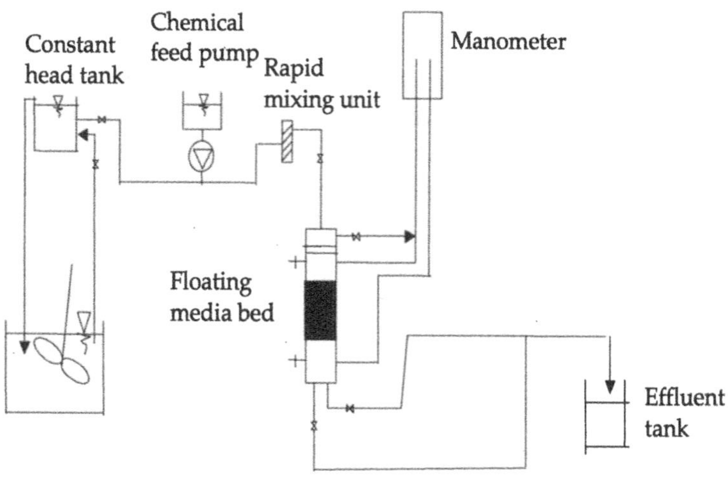

Figure 5.1.1: Schematic of the experimental unit

Results and Discussion

Determination of appropriate coagulant doses

Different coagulant types were examined for synthetic and actual raw water of the water treatment plant using standard jar test. Alum and polyaluminum chloride (PACl) were compared in case of synthetic water whereas actual chemicals used in the operation of water treatment plant, i.e. alum, ferric chloride, PACl, lime and polymer were examined for the latter case. The criteria used to determine appropriate chemical doses were those required to reduce supernatant turbidity to less than 5 NTU while producing large flocculated particle for subsequent removal by floating plastic media pre-filter. From jar test results, it was found that optimum dose of alum and PACl for synthetic water were 30 and 15 mg/l ***(Figure 5.1.2)*** with final turbidity of supernatant of 2.9 and 4.6 NTU respectively. The determination of particle sizes revealed that they were increased from about 5 μm to 75 and 110 μm in flocculated water. Further increase of coagulant dose beyond optimum value increased particle sizes to a maximum value of about 210 μm. Visual observation has shown that PACl gave strongly bonded particles as compared to alum.

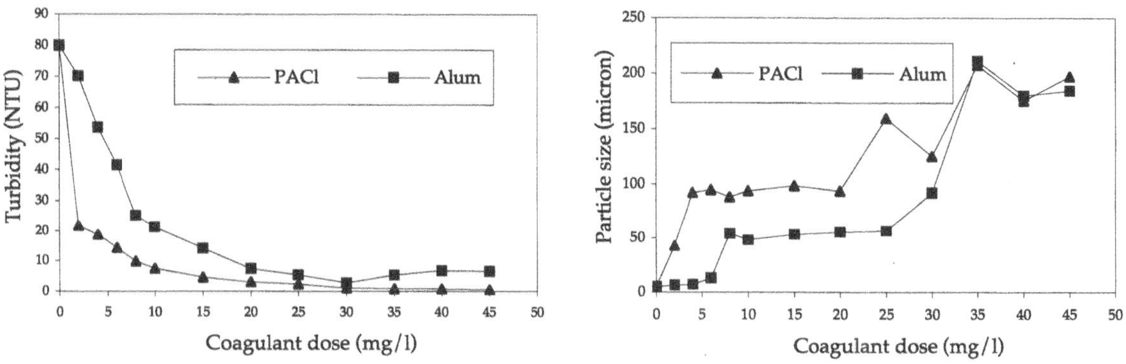

Figure 5.1.2: Effect of coagulant type of dose on effluent turbidity and particle size

The jar test results for raw water of Bangkhen water treatment plant are shown in ***Figure 5.1.3***. In addition to the two different types of coagulant used with synthetic water, a combination of alum, lime and polymer at a dose currently applied in practice at the water treatment plant was also examined. In this case, it was also found that PACl was the best coagulant giving acceptable turbidity level at a minimum dose. The required coagulant doses were 45 mg/l of alum, 35 mg/l of alum/lime/polymer, 20 mg/l of ferric chloride and 10 mg/l of PACl respectively. Determination of size of coagulated water suggested that PACl produced largest flocculated particles with a maximum size of about 400-500 μm.

Particle removal efficiencies in floating plastic media pre-filter

The coagulated water was applied to the experimental system where turbidity and suspended solids removal were evaluated. ***Table 5.1.1*** shows turbidity removal efficiencies in floating plasticmedia pre-filter with different media sizes, bed depths and filtration rates when being applied to synthetic water. It was found that the use of 3 and 6 mm plastic media sizes gave better removal efficiencies than 8 mm, though that of 6 mm size was slightly better than 3 mm. In general, deeper filtration bed and lower filtration rate yielded better turbidity removal efficiency. Highest particle removal efficiency (92%) was obtained with 80 cm bed of 6 mm plastic media and 60 cm bed of mixed size plastic media (3, 6 and 8 mm) at a filtration rate of 5 m^3/m^2-h. High removal efficiencies (of more than 85%) could still be achieved at higher filtration rate (7.5 and 10 m^3/m^2-h) when single media size of 6 mm. was used.

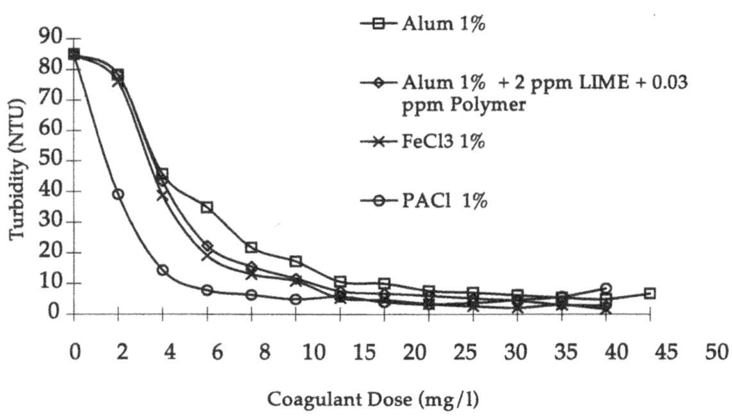

Figure 5.1.3: Jar test results for raw water of Bangkhen water treatment plant

Table 5.1.1: Turbidity removal efficiencies in floating plastic media bed after 6 hours operation

	Removal percentage (%)		
	5 m^3/m^2.h	7.5 m^3/m^2.h	10 m^3/m^2.h
3 mm. media bed			
- 40 cm. depth	56.56	68.91	70.01
- 60 cm. depth	89.23	79.94	76.84
- 80 cm. depth	82.10	90.96	79.38
6 mm. media bed			
- 40 cm. depth	74.84	61.08	64.01
- 60 cm. depth	84.74	89.17	85.61
- 80 cm. depth	92.08	90.71	87.26
8 mm. media bed			
- 40 cm. depth	55.46	54.10	50.56
- 60 cm. depth	80.78	60.15	60.56
- 80 cm. depth	82.91	65.60	72.39
3,6,8 mm multi-layer media bed			
- 60 cm. depth	92.90	70.25	69.42

Accumulation of solids and headloss development in floating plastic media bed

Figure 5.1.4 shows solid concentrations accumulated in floating plastic media bed (in case of 6 mm. uniform media size) at different filtration rates. At lower filtration rate, majority of solid particles was found accumulated at the upper part of filter bed. As the filtration rate increased, accumulation zone shifted downward and accumulated solids were distributed uniformly along the bed. Similar observations were found in other cases (3 and 8 mm media size). The size of particles released from the floating plastic media pre-filter was also found increasing with time ***(Figure 5.1.5)*** and affected by the filtration rate. The particles released from the system at higher filtration rate were found larger than those at lower filtration rate as the accumulation of particles was at higher extent. Also, the particle sizes from uniform size media bed were found larger than those of mixed size media bed which indicated that the agglomeration of particles took place better in single size media bed.

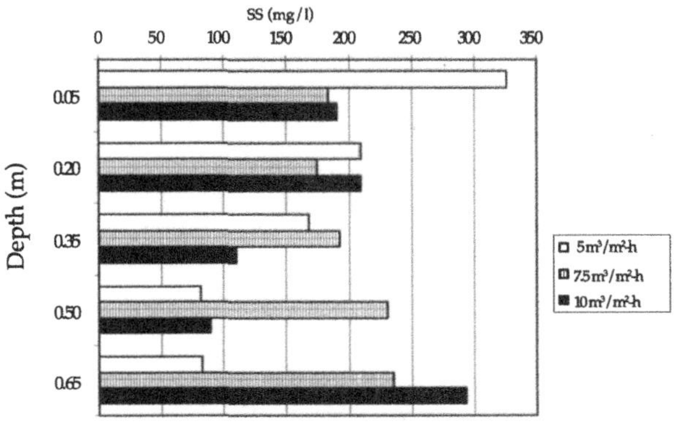

Figure 5.1.4: Accumulation of solids in floating plastic media pre-filter (6 mm size, 0.6 m depth)

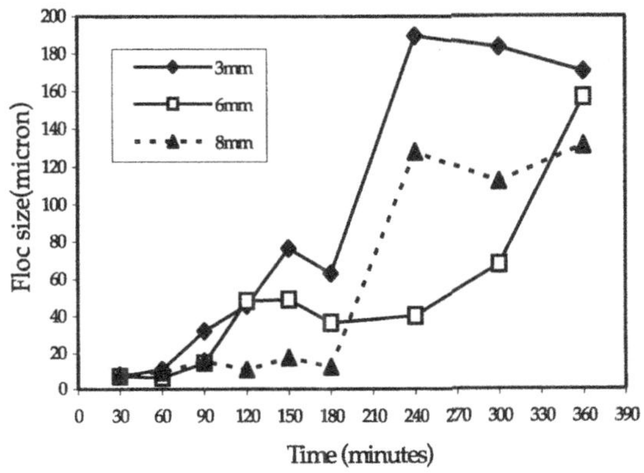

Figure 5.1.5: Particle sizes in the effluent from floating plastic media pre-filter

Headloss development in floating plastic media pre-filter was also governed by media size and filtration rate. The headloss of filter bed were 29, 8.3 and 3.5 cm of water after 6 hours operation using 3, 6 and 8 mm uniform size media at 5 m^3/m^2-h filtration rate. An increase in filtration rate to 7.5 and 10 m^3/m^2-h increased headloss to 60, 4.4, 6.9 cm and 93, 4.5 and 7.8 cm respectively. Headloss in 3 mm media bed was found significantly higher than those of 6 and 8 mm. For mixed size media bed, the headloss fell in range between those of 3 and 6 mm uniform size. Considering from the results in terms of particle agglomeration/entrapment efficiency and headloss development in the system, it was found that 6 mm uniform size media bed was the most appropriate design for the floating plastic media pre-filter in this study.

Similar to typical granular media filtration performance, turbidity removal efficiencies in floating plastic media pre-filter was also related to the ratio between bed depth (L) and media size (d). As shown in ***Figure 5.1.6,*** particle removal efficiencies increased with an increase in L/d irrespective of media size. Considering an acceptable removal efficiency of 80%, it was suggested that the floating plastic media pre-filter should be design at L/d of 100 or more. This is applicable to the system with minimum bed depth of 60 cm and filtration rate of 5 to 10 m^3/m^2-h.

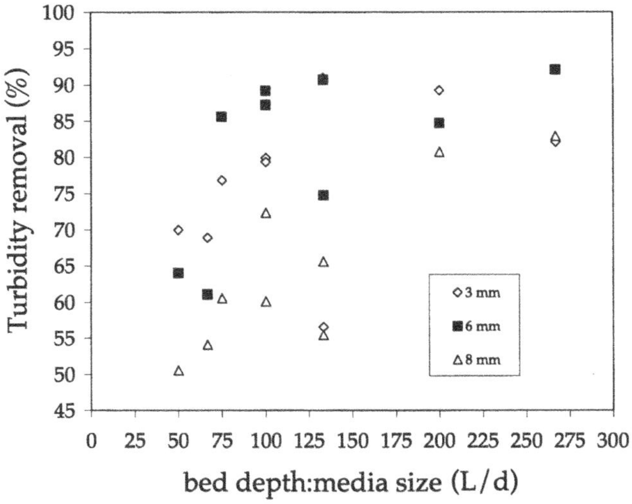

Figure 5.1.6: Relationship between L/d and turbidity removal

Application of floating plastic media pre-filter to actual surface water treatment

The developed floating plastic media pre-filter was subsequently applied to the treatment of raw (surface) water at Bangkhen water treatment plant, Thailand. Two filtration rates, i.e. 5 and 10 m^3/m^2.h, were tested in short term experiment (6 hours). PACl was used as the coagulant at different doses of 20%, 50% and 100% of optimum dose obtained from jar test. ***Figure 5.1.7*** shows the variation of effluent turbidity from the system at different dose of coagulant at each filtration rate. It was found that a coagulant dose of 100% test gave the best result at a filtration rate of 5 m^3/m^2.h whereas 50% dose was optimal for 10 m^3/m^2.h. Average turbidity level in the effluent over 6 hours period at optimum doses were 7.5 and 28.1 NTU and maximum headloss development were 6 and 16 cm of water respectively.

Conclusions

This study was conducted to determine the design criteria of floating plastic media bed to remove turbidity from surface water. It was performed by using synthetic raw water made of kaolin clay particles with constant turbidity level of 80 NTU. From the determination of appropriate coagulant type and dose by using jar test, it was found that polyaluminium chloride (PACl) was better than alum with optimum dose of 15 mg/l. From the filtration by using floating plastic media bed with average size of 3 ,6 ,8 mm and bed depth of 40 , 60 , 80 cm with the filtration rate of 5 , 7.5 , 10 m^3/m^2-h, it was found that the highest turbidity removal efficiency was obtained with 80 cm bed of 6 mm plastic media and 60 cm bed of mixed size plastic media (3, 6 and 8 mm) at a filtration rate of 5 m^3/m^2-h. A deeper filtration bed and lower filtration rate yielded better turbidity removal efficiency. The suspended solids accumulated at the upper zone of filtration bed when low filtration rate was applied. As the filtration rate wasincreased, the accumulation took place evenly throughout the filtration bed. The size of particles released from the floating filter increased with time. Reduction in filtration rate and increase in filter bed depth prolonged its operational period. The appropriate criteria for designing floating media bed for more 80% turbidity removal efficiency was of L/d 100 or more, with bed depth of 60 cm or more, at the filtration rate of 5 to 10 m^3/m^2-h. Application of developed system to actual raw water of water treatment in Bangkok has shown that the system could produce treated water with low turbidity (7.5 NTU) at 5 m^3/m^2-h filtration rate. It could be considered as an efficient pre-treatment unit for conventional sand filter or microfiltration membrane.

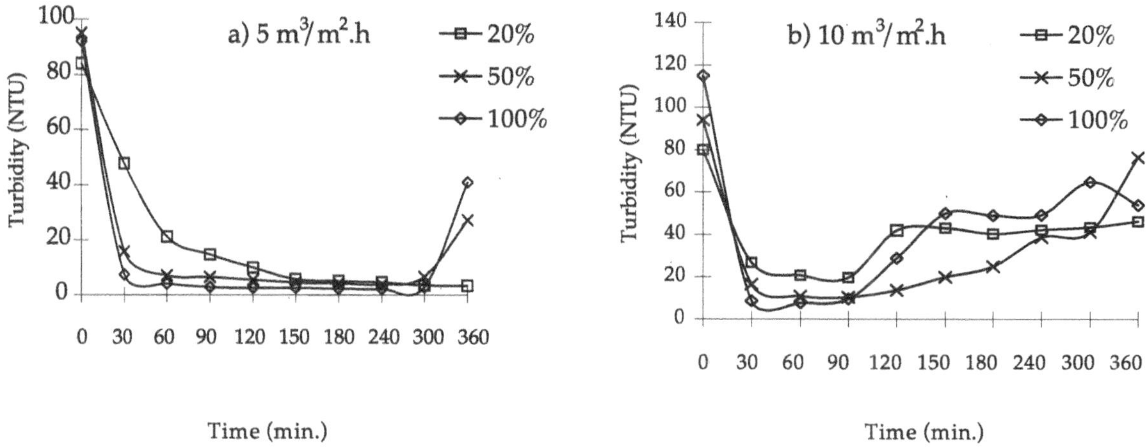

Figure 5.1.6: Variation of effluent turbidity from floating plastic media pre-filter treating raw water of Bangkhen water treatment plant

References

Ben Aim R., Shanoun A., Visvanathan C. and Vigneswaran S. (1993), *New Filtration Media and their Use in Water Treatment*, Proceedings of World Filtration Congress, Nagoya, Japan, 273-276.

Chiemchaisri C., Panchawaranon C., Rutchatanunti S., Kludpiban A., Ngo, H.H. and Vigneswaran S. (2003), Development of Floating Plastic media Filtration System for Water Treatment and Wastewater Reuse, *J. Env. Sci. Health*, **38**, 2359 –2368.

Chiemchaisri C., Chiemchaisri W., Kornboonraksa T., Dumrongsukit C., Threedeach S., Ngo H.H. and Vigneswaran S.(2005), Particle and Microorganism Removal in Floating Plastic Media Coupled with Microfiltration Membrane for Surface Water Treatment, *Wat.Sci.Tech.*, Vol.51, No.10, pp 93-100.

Ngo, H.H. and Vigneswaran S. (1995), Floating Medium Downflow Flocculator with Coarse Sand Filter – A System for Small Community, *Water*, AWWA, July/Aug.1995, 34-37.

Visvanathan C., Werellagama D.R.I.B. and Ben Aim R. (1996), Surface Water Pretreatment using Floating Media Filter, *J. Environmental Eng.* ASCE, January 1996, 25-33.

6 Constructed Wetlands

6.1 Nitrogen removal from wastewater of a fertiliser and chemical manufacturer with the use of two different hybrid constructed wetland systems planted with *Schoenoplectus validus*

Domingos, S[1]. Germain, M[2]. Dallas, S[1].
[1] UNEP-IETC Environmental Technology Centre, Murdoch University, Perth 6150 Western Australia

[2] CSBP Limited, Perth, Western Australia
s.domingos@murodch.edu.au

Abstract

This study analyses the performance of a hybrid Free Water surface/Vertical Flow (FWS/VF) wetland in removing nitrogen from wastewater generated in a fertiliser and chemical manufacturer in Perth, Western Australia during a 14 month period. The constructed wetland currently in use at the fertiliser plant is 12,825 m^2 (135 x 95 m), planted with *Schoenoplectus validus*, has a vertical flow design and has been operated as FWS/VF conjugated system with a hydraulic load = 3.3 cm/d and a retention time of 18 days. Average load removal rates are: 26% for NH_4-N, 52% for NO_3-N and 32% for TN. Parallel to that, an experiment with wetland mesocosms was conducted, this experiment compared the efficiency of a FWS/VF wetland (simulation of the industry's wetland) and an alternative hybrid VF/Horizontal Flow (VF/HF) in removing nitrogen from the industry's wastewater. Systems were planted with *Schoenoplectus validus*. The performance in nitrogen removal for the FWS/VF mesocosm was superior to the actual wetland's performance. The best nitrogen removal rate was achieved by the VF/HF system, with the first VF stage alone removing 82.9% of ammonia in 2.88 days while the FWS/VF removed 78.2% of the ammonia in 10.8 days.

Key Words

Constructed wetlands; hybrid systems; vertical flow; horizontal flow; nitrification; industrial wastewater.

Introduction

Westfarmer's CSBP manufacturing plant in Kwinana, Western Australia, produces Ammonia, Ammonium Nitrate, Nitric Acid, Compound Nitrogen and Phosphorous Fertilisers, Superphosphate, and Chlor-alkali products. As part of their operation the plant generates different types and amounts of effluent, the main sources are cooling tower blow-down water, combined wastewater pumped from production facilities and stormwater runoff. Since constructed wetlands (CW) have proven to be a cost-effective alternative in wastewater treatment, CSBP built a constructed wetland on-site and started monitoring it in 2004. Wastewater originates at different locations in the plant and is firstly pumped into a containment tank and from there to the constructed wetland. The main concern of the effluent is its inorganic nitrogen contents, primarily in the form of NH_4-N and NO_3-N.

The main NH_4^+ removal pathway in a wetland is nitrification since volatilization is insignificant at a pH below 8 (Reedy and D'angelo, 1997) and adsorption to organic sediments only occurs in the early stage of a wetland's life (Kadlec and Knight, 1996).

According to Kadlec and Knight (1996) the main NO_3- removal pathways in wetlands are plant uptake and denitrification. Nitrogen plant uptake capacity of emergent macrophytes within reed beds ranges between 200 to 2500 kg $ha^{-1}year^{-1}$, but plant uptake is only of quantitative importance in low loaded systems (Brix, 1994), denitrification is likely to play a major role in removing NO_3- from overloaded systems. Reed *et al.* (1995) report plant uptake accounting for less than 10% of the nitrogen removed in two systems where plants were harvested.

This paper reports on the results of a 14 month operation analysis for CSBP pilot constructed wetland cell and on the preliminary results from pilot scale hybrid Vertical Flow/Horizontal Flow and Free Water surface/Vertical Flow systems.

Methods

CSBP wetland: performance analysis

CSBP's pilot constructed wetland characteristics are: 135 m long x 95 m wide x 1 m deep; (~ 1.3 ha) it has a vertical flow design; the medium is sand with a void ratio of approximately 0.3. The influent distribution pipe is laid on the surface of the sediment and the treated effluent drainage pipes are laid underneath the sand column on the surface of the liner, drainage pipes are covered with rocks and geotextile fabric (fig. 1). The wetland vegetation consists of River Club-rush (*Schoenoplectus validus*). CSBP still has 3 ha of land available for the construction of more wetland cells.

The wetland is operated as a hybrid Free Water Surface / Vertical Flow (FWS/VF) system. The medium is kept constantly saturated representing the VF and there is a 0.3 m water column on top of the sediment which represents the FWS. It was designed to receive an effluent flow of 2653m^3/day with the following average nitrogen concentrations: NH_4-N 22mg//L, NO_3-N 13 mg/L, TN 35 mg/L. The target nitrogen reduction was 50% but such reduction has not been achieved with the current wetland operational scheme. Up to date (Aug/04 – Feb/06) data show that the wetland receives an average effluent volume of 421 m3/day, which corresponds to a Hydraulic Loading Rate (HLR) of 3.28 cm/day, with average nitrogen concentrations NH_4-N= 36 mg//L, NO_3-N=12 mg/L, total nitrogen (TN)=48 mg/L. The average Hydraulic Retention Time (HRT) for the period was 18.3 days.

The wetland is to our knowledge the best monitored system in Western Australia. It is equipped with flow meters and automated samplers so every time the wetland is filled or drained samples are collected from the inlet and outlet pipes. Samples are analysed for pH, conductivity, ammonia, nitrate, total nitrogen, total phosphorous and a wide range of metals in CSBP Kwinana laboratory. In this paper we will focus on nitrogen forms which are the contaminants of concern due to their high concentration. Data from 14 months of operation was analysed, starting in September 2004 and finishing in October 2005.

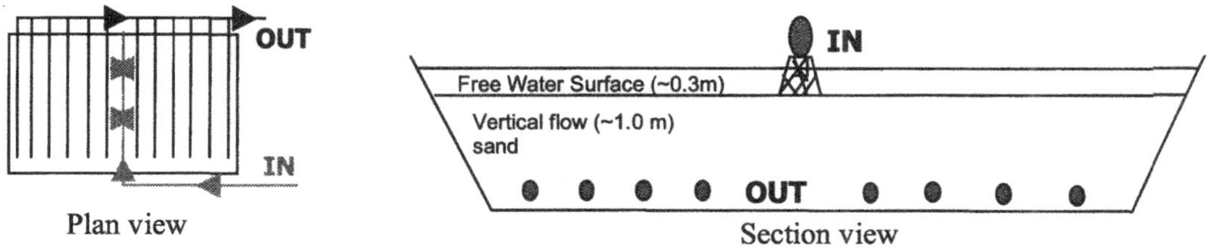

Figure 6.1.1: Design of treatment wetland in CSBP fertiliser and chemical manufacturer, Kwinana, WA, Australia.

Experimental Set Up

In order to test different performances of hybrid systems in removing nitrogen from the wastewater generated at CSBP Kwinana experimental systems were set up at the Environmental Technology Centre, Murdoch University, Western Australia. All system were planted with the common river club rush *Schoenoplectus validus* and set up in duplicates. The effluent to be treated consisted of industrial wastewater generated at Wesfarmer's CSBP fertiliser production plant; such effluent previously passed through a containment tank and then flowed to the CW systems. During the experiment wastewater was brought from the industry to the location of the experiment every two weeks and stored in the dark at an average temperature of 13.3 °C (average air temperature for the period). Water analyses made by the time of collection at the industry and after it had been stored at location of the experiment for two weeks showed that the concentration of ammonia, nitrate and total nitrogen did not vary significantly.

Once planted, experimental wetlands were left "resting" for approximately a month, necessary for vegetation and bacteriological establishment, during this time plants were irrigated with scheme water for 2 weeks and then with the wastewater, this was considered enough to start up the biological process. After this time, sampling started.

Free Water Surface/Vertical Flow hybrid system (FWS/VF)

This set up was meant to simulate in a smaller scale the design and operation mode of CSBP's existing hybrid wetland. For such simulation 200L plastic barrels were used. The system was fed manually with the use of a measured 10L bucket. For the outflow arrangement, PVC perforated pipes were placed on the bottom of the barrels, to avoid clogging, these pipes were covered by a 10 cm layer of gravel (size 1 to 2 cm) and then a 50 cm layer of sand. The depth of the bed was 0.6m. In order to provide a carbon source for denitrification, mulch was placed in one layer within the sand columns. The sand media was kept totally saturated and the water level was maintained 0.2 m above the sand, the 0.6 m saturated sand corresponds to the VF system and the water column of 0.2 m represents the FWS system. The operation mode of the system followed CSBP's wetland fill and drain operation. On every 6th day the systems were batch loaded with 18 litres of effluent. Before batch loading the system was drained until the water level reached the sand surface, samples were taken at the time of draining. Hydraulic Retention Time (HRT) was designed to be 12 days in total, with the water remaining 6 days in the FWS component and 6 days in the VF component. ***(Figure 6.1.2-A)***

Vertical Flow/Horizontal Flow hybrid system (VF/HF)

For the VF system the same design of the FWS/VF system was utilised but in this case the water level was maintained just below the surface of the sand (Figure 6.1.2-B). The system was batch loaded manually with the use of a measured 10 L bucket. There was an air pipe with slots near the bottom end placed vertically through the media which functioned as a passive air pump based on a fill and draw sequence (Reed *et al.*, 1995; Cooper, 1996; Green *et al.*, 1997). As water drains from the bed fresh air is sucked into the lower layers of the bed trough the pipes and as the bed is filled with water oxygen depleted air will be pushed out of the bed through the aeration pipes. The HRT for the VF component was designed to be 2 days, after 2 days the systems were allowed to rest for two days so oxygen would penetrate into the sediment. The volume of effluent for each VF wetland was 22 L.

The HF systems consisted of 1.5m x 0.5 m x 0.4m (length x width x depth) tubs. The bottom had a 1.0% slope to assist the water flow through the system. The media chosen was gravel (size 1-2 cm) due to its higher hydraulic conductivity, the depth of the gravel bed was 20 cm. For the inlet a cross sectional PVC pipe with orifices was laid at the proximal end of the tubs (Figure 6.1.2-C). For the outflow arrangement, each tub had a collecting PVC perforated pipe placed at its distal end; this outlet pipe had a height control mechanism which allowed controlling the water level within the vegetated bed. After leaving the experimental tubs, the treated water was sampled and discharged to the sewerage.

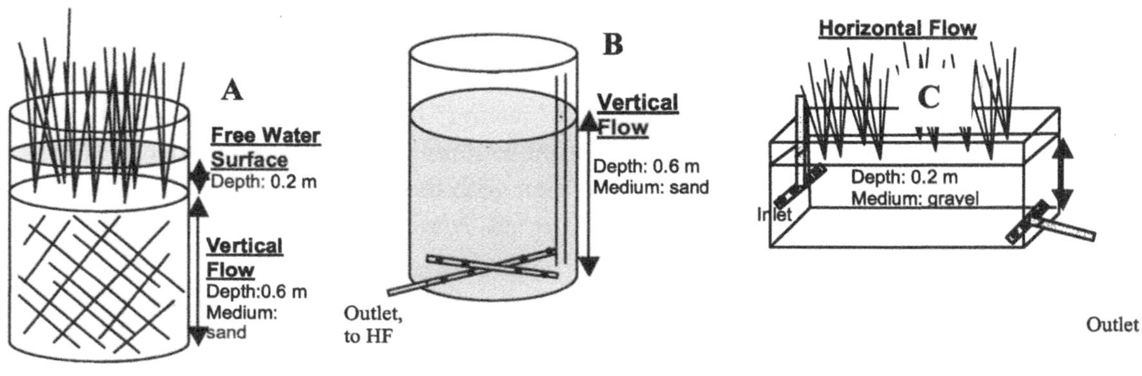

Figure 6.1.2: A- Free Water Surface/Vertical flow wetland. B- Vertical Flow component (plants have been omitted). C- Horizontal Flow component.

Results and Discussion

CSBP wetland 14 month analyses

Total volume of wastewater pumped in and out of the wetland during the 14 month period was 175,724 m^3 and 158,586 m^3 respectively, total in/out-flow for each month can be seen in ***figure 6.1.3***. This volume corresponds to an average flow of 412.5 m^3/day and a HLR of 3.21 cm/day.

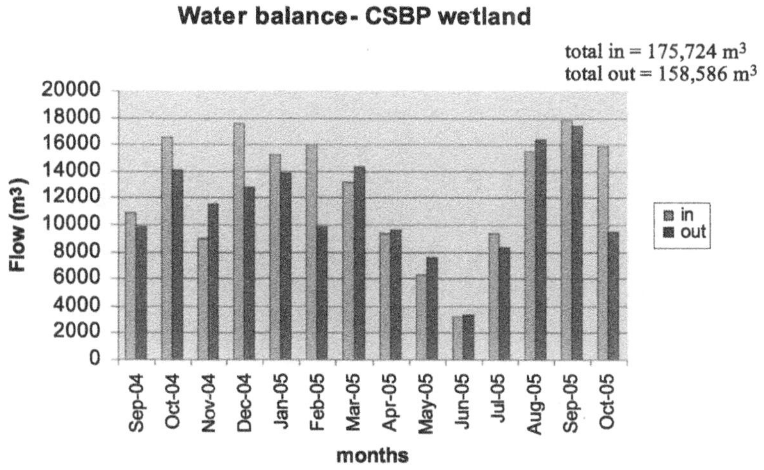

Figure 6.1.3: Total volume of wastewater pumped into and out of CSBP's wetland from September 2004 to October 2005.

Ammonia, nitrate and total nitrogen mass balance values can be seen in figures 6.1.4, 6.1.5 and 6.1.6 respectively.

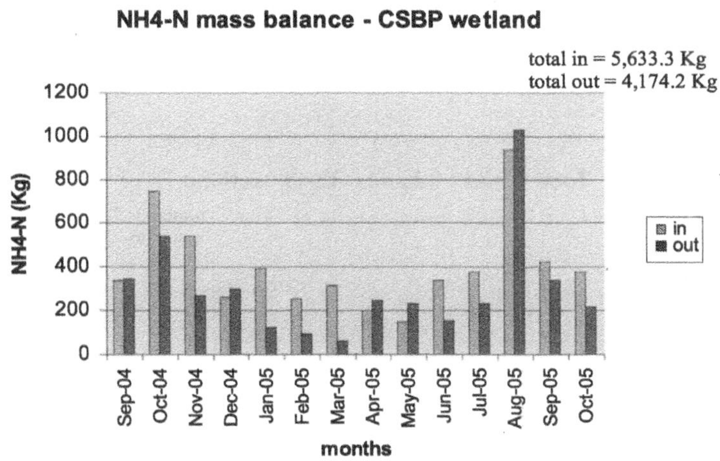

Figure 6.1.4: Total mass of NH_4-N pumped into and out of CSBP's wetland from September 2004 to October 2005.

During the period analysed 1459 Kg of ammonia were removed from the wastewater, this represents a load removal of 25.9%. In terms of concentration the average NH_4-N in and out were 34.9 mg/L and 29.9 mg/L.

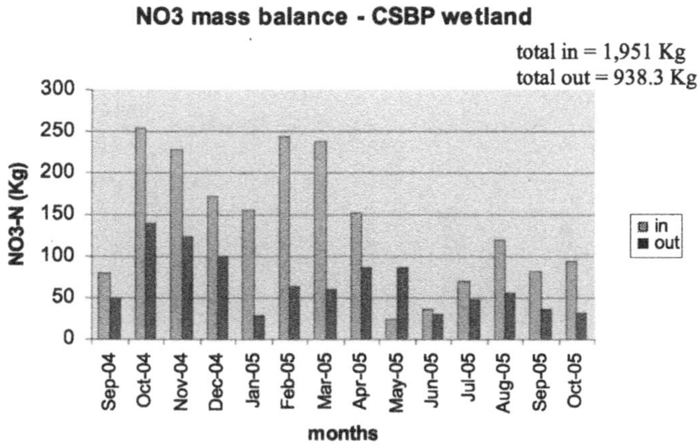

Figure 6.1.5: Total mass of NO_3-N pumped into and out of CSBP's wetland from September 2004 to October 2005.

The nitrate load removal (51.9% or 1012 Kg) was better when compared to the load removal of ammonia; this can be explained by the predominance of anaerobic conditions in the wetland as it has been recently verified through dissolved oxygen (DO) measures in the surface and subsurface components of the wetland (full details of this part of the study to be described elsewhere). Although presenting high DO values in the free water surface component, average of 7.5mg/L, DO dropped to an average of 0.4mg/L in 50 cm depth of the subsurface component. These findings are in agreement with the operation mode which never drains the system completely and therefore does not allow for air to be transferred into the substrate, favouring anaerobic conditions in the subsurface.

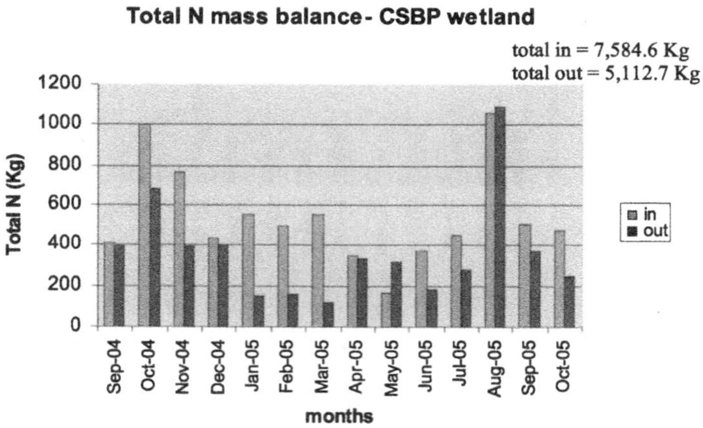

Figure 6.1.6: Total mass of TN pumped into and out of CSBP's wetland from September 2004 to October 2005.

The removal of total nitrogen by the wetland accounted for 2,471.9 Kg which correspond to 32% of the initial load. Considering the long HRT of about 18 days ammonia and TN values in the outflow were expected to be lower. When the wetland was designed in 2003 the target removal for TN was 50% but so far such removal has only been achieved in some months, without any consistency. Anderson *et al.* (2005) reported on FWS wetlands for wastewater treatment in Sweden removing from 24% of TN in 6-7 days to 68% of TN in 14 days.

Oxygen transfer into the wetland substrate might be the main limiting factor for nitrification to occur and thus further removal of ammonia and TN (since most of the nitrogen is in the ammonia form), however, other factors known to affect nitrification such as heavy metal accumulation (Cela and Summer, 2002) and high and fluctuating salinity (Yu *et al.*, 2002) have to be further investigated.

The innovative design of the CSBP wetland has proven to be efficient in nitrate removal but inefficient to remove ammonia.

Experimental Set Up

Sampling started in June 2006 and finished in August 2006. Influent NH_4-N concentration ranged from 30 to 68 mg/L, NO_3-N concentration from 5.33 to 25.37 mg/L and TN from 35.3 to 93.4 mg/L. The mean values for influent nitrogen forms, hydraulic retention times (HRT) and effluent values can be seen in Table 6.1.1.

Table 6.1.1: Treatment performance summary from experimental wetlands at the Environmental Technology Centre.

	AREA m^2	HLR cm/d	HRT days	NH_4-N mg/L	Reduction %	NO_3-N mg/L	Reduction %	TN mg/L	Reduction %
Influent	-	-	-	41.50	-	12.88	-	54.38	-
VF	0.23	3.32	2.88	7.08	82.9	24.24	-88.2	31.31	42.4
FWS/VF	0.23	1.44	10.83	9.04	78.2	4.96	61.5	13.99	74.3
VF/HF	0.98	0.27	8.25	2.19	94.7	10.03	22.1	12.22	77.5

Note: HLR, HRT and concentrations are mean values. VF refers to first stage only of hybrid VF/HF system.

It is clear that ammonia is very efficiently converted into nitrate within the vertical flow (VF) system, which is the first stage of the hybrid VF/HF, but this system alone is not enough to remove nitrate, on the contrary, nitrate concentration nearly doubles in the system. Nitrate removal has to be achieved in a subsequent system, here represented by the HF bed. The VF system alone removed 82.9% of ammonia in 2.88 days HRT while the FWS/VF system removed 78.2% in 10.83 days HRT. Results obtained here are in agreement with other experiments using VF wetlands, where due to good oxygen transfer, nitrification is the major mechanism for ammonia removal (Green *et al.*, 1997 Laber *et al.*, 1997, Cooper, 2005).

Nitrate is very well removed by the FWS/VF system which simulates the CSBP wetland, and this finding matches the current performance of the wetland located at CSBP. Total nitrogen removal by this system was intermediate between the VF system alone and the combined VF followed by HF, even though it had the longest retention time amongst all the treatments.

The hybrid VF/HF system has had the best performance in terms of total nitrogen and ammonia removal to date, however it has not been successful in removing nitrate. This can be explained by the fact that the HF system was shallow (only 20 cm deep) and had been operated in such a way that allowed oxygen to infiltrate into the gravel media, as half of its volume was drained prior to receiving the nitrified effluent from the VF system. Such operation mode changed to the HF bed being kept constantly saturated. Another factor which may have limited nitrate reduction in the HF was the lack of a carbon source to support denitrification.

Of note also is the relatively high performance of the CSBP model (FWS / VF) compared to the actual wetland's performance, this can be explained by the fact that the experimental wetland is small and would have a greater and positive edge effect affecting removal processes, other factors could be heavy metals accumulation within the two year old CSBP wetland and also greater salinity fluctuations.

Conclusions

Results from this experiment are in accordance with what other authors have reported for VF wetlands. While it may be premature to draw conclusions at this stage the results are certainly promising that increased nitrification can be achieved with vertical flow wetlands, this might be seen an option to the industry if further ammonia and TN removal rates are to be achieved. Further research testing varying HLR, HRT as well as resting time of the VF bed is being carried out and will allow for more consistent designs and operation modes of hybrid wetland systems for industrial wastewater treatment.

References

Anderson, J. L., Bastviken, S. K. and Tonderski, K. S. (2005). Free water surface wetlands for wastewater treatment in Sweden: nitrogen and phosphorous removal. Water Science and Technology. **51** (9). p39-46.

Brix, H. (1994). Functions of macrophytes in constucted wetlands. Water Science and Technolgy. **29** (4). p71-78.

Cela, S. and Summer, M. E. (2002). Critical concentrations of copper, nickel, lead and cadmium is soils based on nitrification. Commun. Soil Sci. Plant Anal. **33** (1&2) p19-30.

Cooper, P. (2005). The performance of vertical flow constructed wetland systems with special reference to the significance of oxygen transfer and hydraulic loading rates. Water Science and Technology. **51** (9). p.81-90.

Cooper, P. F., Jobk G.D., Green, M. B. and Shutes, R. B. E. (1996). Reed Beds and Constructed Wetlands for Wastewater Treatment. Wrc Publications, Medmenhan, Marlow, UK.

Green, M., Friendler, E., Ruskol Y. and Safrai, I. (1997). Investigation of alternative method for nitrification in constructed wetlands. Water Science and Technology. **35** (5). p63-70.

Kadlec, R. H. and Knight, R. L. (1996). Treatment Wetlands. Lewis Publishers, Boca Raton, Fl.

Laber, J., Perfler, R. and Haberl, R. (1997). Two strategies for advanced nitrogen elimination in vertical flow constructed wetlands. Water Science and Technology. **35** (5). p71-77.

Reed, S. C., Crites, R. W. and Middlebrooks, E. J. (1995). Natural systems for waste management and treatment. 2nd Edition. Mc Graw Hill.

Reedy, K. R. and D'angelo, E. M. (1997). Biochemical indicators to evaluate pollutant removal efficiency in constructed wetlands. Water Science and Technology **35** (5), p 1-10.

Yu, S.M., Leung, W.Y., Ho, K.M., Greenfield P.F. and Eckenfelder W.W. (2002). The impact of sea water flushing on biological nitrification-denitrification activated sludge sewage treatment process. Water Science & Technology **46** (11-12). p209–216

6.2 Constructed Wetlands for Wastewater Treatment in Western Australia

Stewart Dallas[1], Anda, M, Ho, G
[1] Environmental Technology Centre, Murdoch University, Perth 6150 Western Australia
s.dallas@murodch.edu.au

Abstract

The potential for constructed wetlands (CWs) in Western Australia is enormous as they can provide cost-effective wastewater treatment systems with an effluent quality suitable for reuse which are simple and inexpensive to operate and maintain when compared to conventional treatment systems. As a result they represent an appropriate and sustainable technology for wastewater treatment – properties which have been widely documented. The largest drawback is the area they require which can make their application to urban areas problematic; however they are particularly well suited to small towns and rural communities where sufficient land is more likely to be available. In addition regulatory bodies are currently preoccupied with the potential for mosquito habitat which can be avoided with correct design.

Constructed wetlands can be considered to be of three principal types: horizontal subsurface flow (SSF), vertical flow (VF) and free water surface (FWS). The design of SSF and FWS wetlands historically has been based on data from numerous wetlands worldwide which has been used to derive empirical equations – the 'black box' approach. The Environmental Technology Centre is now beginning to monitor several different types of constructed wetlands in southwest Western Australia and data more relevant to the local climate is being compiled. This includes different wastewater types including greywater, levels of treatment, suitable plant species, local climatic effects and general construction details. This monitoring program will ultimately lead to improved wetland design for Western Australia conditions.

Key Words

Australia; constructed wetlands; wastewater

Introduction

Constructed wetlands are highly complex ecological systems that separate and transform contaminants by physical, chemical and biological mechanisms and an understanding of these mechanisms is essential if the development of models for predicting pollutant removal is to be achieved (Crites and Tchobanoglous, 1998; USEPA, 1999). It has been recognised that while in a qualitative sense these processes are known, it is only in a few cases that sufficient and adequate measurements have been made to provide a more quantitative assessment. A large part of the difficulty has been separating contaminant removal from transformation processes as both occur simultaneously, and further complicated by the fact that contaminants may be present in several forms. As a result constructed wetlands have been treated, for design purposes, as black boxes.

The aim of this paper is to review the design of constructed wetlands for wastewater treatment and strongly suggest the collection of data for design under Australian characteristic conditions.

Modelling of Constructed Wetlands

The design of constructed wetlands for wastewater treatment worldwide has been by and large based upon the models and findings of either Kadlec and Knight (1996) or Reed *et al.* (1995), and the assumption of first order plug flow. In both cases rate constants have been derived from the results of pre-existing wetlands. These two principal models are differentiated by the choice of rate constant - either volumetric and temperature dependent (Reed model) or area based and independent of temperature (Kadlec & Knight model). First-order models presume ideal plug-flow behaviour with steady-state conditions for flow and concentrations and predict an exponential concentration profile between inlet and outlet (Rousseau *et al.*, 2004). These models only consider free water surface (FWS) and subsurface flow (SSF) constructed wetlands, or reedbeds, and not vertical flow (VF) wetlands. This is presumably due to the paucity of vertical flow wetlands and hence data, as their development has only occurred in the last ten years or so. This appears to be despite the fact that preliminary performance data suggests vertical flow systems are capable of superior performance in terms of biological oxygen demand (BOD), total suspended solids (TSS), total nitrogen (TN) and total phosphorus (TP) removal (DLWC, 1998). The use of Reed's or Kadlec & Knights' models would overestimate the area required for vertical flow wetlands as these models are derived from FWS and SSF systems.

The Kadlec & Knight model has been derived from analysing the performance of all the FWS and SSF constructed wetlands in the North American Data Base (NADB). Reed's model is similarly derived from a smaller set of performance results although this is not entirely clear (DLWC, 1998). Both models are based upon designing according to the pollutant which requires the largest land area for its removal and using this as the limiting design factor. It has been noted that wetlands in the NADB were not in general designed using the first order plug flow assumption and further because of this 'black box' approach to design, "any given system is likely to be over-designed for the limiting pollutant and therefore by definition, must be over-designed for all other pollutants" (DLWC, 1998).

The NADB (Version 1) was prepared with funding from the USEPA and released in 1994 and contains information on 174 natural and constructed treatment wetlands in over 30 US states and Canadian provinces (USEPA, 1999). It contains system specific information such as dimensions, flow rates and plant species in addition to water quality data but the data, in the words of the USEPA, "is not of uniform quantity and quality, which makes it inappropriate for design or modelling purposes". Version 2 has subsequently been prepared and contains information on treatment wetlands at over 245 locations in the US and Canada, but again "the data is not adequate for design or modelling" (USEPA, 1999).

First-order plug-flow models

The general form of the equation for first-order plug-flow models is:

$$\ln(C_o/C_i) = -kt \quad (1.1)$$

where: Ci = influent pollutant concentration (mg/L)

Co = outlet effluent pollutant concentration (mg/L)

k = reaction rate constant

t = hydraulic residence time (days)

For Reed's model this equation is used to calculate BOD, ammonia (NH_3) and nitrate (NO_3) removal only with the following rate constants for SSF wetlands at 20°C respectively: 0.678 day^{-1}, 0.2187 day^{-1}, 1.000 day^{-1}. Separate equations are used for TSS and total phosphorus removal, and pathogen removal is based on the same approach as is used for stabilisation ponds as follows:

$$C_o/C_i = (1 + kt)^{-n} \quad \text{where n = number of cells in series.} \quad (1.2)$$

For Kadlec and Knight's model this equation is modified to include a background pollutant concentration (C*) and the equation is used to model all pollutants, and they conclude that this is in general independent of temperature except for nitrogen removal. The general form of the equation is as follows:

$$\ln[(Co - C^*)/(C_i - C^*)] = -k/q \quad (1.3)$$

where: C* = background pollutant concentration (mg/L)

q = hydraulic loading rate (m/year)

The corresponding values at 20°C of k and C* for BOD and fecal coliform removal respectively are: 180 m/year, 3.5+0.053C_i ; 95 m/year, 10 cfu/100 mL.

The effect of temperature on k_v is generally accounted for through a modified Arrhenius temperature dependence, as summarised by equation 1.4:

$$k_v = k_{v20}\, \theta^{(T-20)} \quad (1.4)$$

where k_{v20} = first-order volumetric reaction rate coefficient at 20° C, d^{-1}

θ = temperature correction factor

T = water temperature, °C

Wetland hydrology

In a treatment wetland where there are generally no stormwater or groundwater inflows/outflows, rainfall and evapotranspiration (ET) are only other inputs/outputs to a continuous 'wastewater' inflow. ET occurs with strong diurnal and seasonal cycles because it is driven by solar radiation which in turn is cyclical by nature. Thus ET can be an important water loss on a periodic basis. In order to accurately quantify the overall water budget for a given wetland system the following equation applies:

$$Q_i - Q_o + PA - ETA = dV/dt \qquad (1.5)$$

Where: Q_i = inflow rate (m^3/day),

Q_o = outflow rate,

P = precipitation rate (m/d),

ETA = evapotranspiration + evaporation rate (m/d)

A = wetland top surface area (m^2)

Typically Q_o is not monitored due to the inherent difficulty of doing so and this water budget formula is used to determine the overall mass balance. For most design purposes historical monthly average precipitation is sufficient (Kadlec & Knight, 1996) with estimated rates of ETA based on a factor times the pan evaporation. For the determination of rate constants however it is necessary to know the outflow rate Q_o.

Application of the models for Australian conditions

The USEPA (1999) go to some length to describe the popular misconception that wetland design is well-characterised by published design equations. Despite the not insignificant job of creating the NADB undertaken by the USEPA, they are quick to point out the limitations of using this database for wetland design. They cite a lack of quality data of sufficient detail, both temporally and spatially; the combining of data from wetlands with rigorous control (QC) with data from wetlands with randomly collected data with little QC; the combining of data from small wetlands with minimal pretreatment with data from large wetlands used for polishing secondary effluent; lack of paired influent-effluent samples, grab samples instead of composite samples and various other shortcomings such as limited or non-existent site-specific temperature and precipitation data. They conclude that "the resulting data combinations...have sometimes been used to create regression equations of questionable value for use in design".

Herein lies the crunch–is the continued reliance on regression-derived treatment kinetics/models based upon questionable raw data from North American treatment wetlands a sensible basis upon which to design treatment wetlands for Western Australia? Others have already echoed this sentiment with the DLWC (1998) saying that a far better approach would be to develop a national database for Australian conditions and this has been echoed by Headley (Headley and Davison, 2003).

Other variables which impact upon the suitability of these models include hydrological influences, influent concentrations, hydraulic retention time, void fraction and water depth. Kadlec and Knight (1996) comment that there is at present no way to distinguish between the two models (theirs and Reed's) as there is currently insufficient understanding of the effect of saturated media depth.

An alternative approach to the two models described above is the 'sequential model' developed by Gearheart and Finney (USEPA, 1999) for free water surface wetlands. This model is based upon the fact that while some of the pollutant removal mechanisms are to some degree occurring simultaneously throughout the wetland, the majority of the removal occurs in a sequential fashion. It is well recognised that TSS and BOD removal for example occur in the initial portion of the wetland. The implication is that, if for example, the majority of pollutant removal occurs in the first 30% of the wetland then the rate constant as determined by using the complete system, and retention time, is significantly down-rated. It is this type of problem that Crites and Tchobanoglous (1998) say "pervades most of the information reported in the literature on constructed wetlands, where only input and output values are used in determining the apparent removal-rate constant". The sequential model indicates three sequential compartments: Zone 1 – a vegetated initial section of two days retention time (maximum monthly flow) for the bulk of the flocculation and sedimentation; followed by Zone 2 – an aerobic open water section of two days retention time where BOD reduction and nitrification can occur and finally; Zone 3 - a vegetated polishing section of two days retention time for further reductions in TSS and nitrogen via denitrification. A complete description of this model is given in USEPA (1999).

While it is generally agreed that temperature has an effect on the rate of nitrogen removal there is considerable debate about whether the same is true for BOD (Reed *et al*, 1995). The implication of temperature dependence is that treatment performance may vary with season and climate. This is significant and has implications for the design of constructed wetlands in Australia where average water temperatures are generally higher than those in much of the United States, and from where the NADB was derived. It has been speculated that this raises the potential for overdosing if Kadlec and Knights' approach is adopted as the actual performance may be more than predicted (DLWC,1998). In effect, the temperature correction factor is a "fudge-factor" that causes the rate of reaction predicted by the model to either increase ($\theta>1$) or decrease ($\theta<1$) with temperature (Headley and Davison, 2003).

Conducting a mass balance is the most accurate way to characterise water quality functions in a wetland as it allows closure for the chemical of interest to be achieved, but as the outflow volume is often not monitored this is rarely achieved (Kadlec and Knight, 1996). Measurements of chemical composition (eg in mg/L) of wetland flows are the most obvious method of characterising water quality functions however, such measurements by themselves can be very misleading. Hence the need for a water budget and mass balance. Ideally closure should be demonstrated where possible for the chemical constituent concerned but "unfortunately closure has rarely been demonstrated for any chemical in any wetland" (Kadlec and Knight, 1996).

To achieve closure it is necessary to use flow-weighted concentrations with time-average samplers (i.e flow-proportional samplers). In the event the flow rate, or concentration, is constant then there is no difference between flow-weighted averages and time averages. This difficulty of synchronous sampling may be alleviated in the mass balancing process by selecting a mass balance period that spans several detention times (Crites and Tchobanoglous, 1998).

The work conducted by Headley and Davison (2003) is perhaps the most important published research to investigate rate coefficients for Australian conditions. In this study based in north-eastern NSW data from 13 reedbeds over 8 years was analysed in an attempt to establish model parameters for the prediction of BOD and total nitrogen (TN). The principal findings were volumetric reaction rate coefficients (k_{v20}) of 0.52 d^{-1} and 0.18 d^{-1} which gave reliable predictions of BOD and TN removal respectively for reed beds greater than six months old treating combined domestic wastewater, greywater or laundry wastewater. TN removal was found to be independent of temperature over the experimental range of 14.8°C to 25°C while a temperature correction factor (θ) of 0.95 was found to apply for BOD removal indicating a decrease in performance with increasing temperature. Background concentrations (C*) were found to be 5 mg/L and 1.5 mg/L for BOD and TN respectively. Of interest was the finding that wastewater type was an important variable. School wastewater for example was found to have a substantially lower reaction rate coefficient (0.52 d^{-1}) for TN removal. This was attributed to the greater proportion of nitrogen-rich urine and higher BOD likely at a day-use only facility as compared to a typical domestic situation where dilution with greater volumes of greywater takes place.

The first-order plug-flow model has been widely used for stormwater wetland design in Australia and yet the "assumption of constancy of the model parameters – decay rate and background concentration is not really the case" (Wong *et al.*, 2006).

The situation in Western Australia

Very few constructed wetlands for wastewater treatment exist at present in Western Australia. Perhaps the first constructed wetland to the author's knowledge was the Mundaring wetland designed by a local consulting firm for final nutrient polishing of secondary treated municipal wastewater (Turner *et al.*, 1996). This free water surface (FWS) wetland forms part of a Water Corporation wastewater treatment plant and was designed to receive 100m^3/day at a flow rate of 6L/sec and an average hydraulic loading of 0.05L/m^2.day. Final design effluent characteristics were TSS < 5mg/L, BOD < 5mg/L, total nitrogen < 10mg/L and total phosphorus < 1mg/L with the effluent reused for irrigation of vegetated public areas. The wetland contains several berms (or baffles) and varying water depths to maximise the flow path and biological function respectively. The selection of plant species was based on their local occurrence in the region and proven performance in wastewater wetlands and consisted of the emergents *Schoenoplectus validus* and *Baumea rubiginosa* and submergents *Triglochin procera* and *Potamogeton pectinatus* for maximum nutrient uptake. No data on the wetland's performance has been published although increased bacteria levels and algae as a result of bird droppings have apparently been of concern. Construction costs for the wetland were also significantly higher than envisaged due to the presence of significant rock in addition to the fall of some three metres across the site. This significantly impacted the wetland, cost effectiveness and could have been avoided with improved site investigations prior to tender award.

This is currently the only wetland on which details of design have been published although several wetlands for abattoir effluent are known to exist in some rural areas. A natural wetland in Derby Western Australia has also recently been partially converted for effluent polishing although no water quality data is available.

Recent attempts to establish constructed wetlands in the Mandurah area south of Perth for small and large scale domestic greywater treatment have been subject to significant investigation at the planning stage by the regulatory authorities (Anda, 2005). The principal concerns have been the potential for mosquito habitat and the predicted levels of phosphorus removal. The concern for mosquito habitat has meant that any constructed wetlands planned in the region for the foreseeable future will, in order to achieve approval, have to be of subsurface flow design with a mosquito monitoring program in place. This despite the fact that "by definition, subsurface-flow wetlands are not mosquito habitats, unless overextended and with surface pooling" (Russell, 1999). It is possible that this requirement in hand with the increased cost of subsurface as compared to free water wetlands may reduce the uptake of constructed wetlands generally.

Conclusion

The uptake of constructed wetlands for wastewater treatment in Western Australia has been hampered by the perceived mosquito risk and the subsequent requirement for subsurface flow only wetlands. It is likely that this requirement in hand with the increased cost of subsurface as compared to free water wetlands may reduce the uptake of constructed wetlands generally. Further, the application of wetlands in Western Australia has been largely for polishing purposes only. This has not proven entirely successful or economically viable and does not reflect their true potential. As has been demonstrated in Western Europe for example many hundreds of constructed wetlands are currently successfully treating not only secondary but primary effluent.

Further research is required as are full-scale demonstration sites at which the full potential of all three types of constructed wetlands can be confirmed. In particular more research on vertical flow (VF) wetlands is needed due to their compact size, ability to remove nitrogen and the fact that no suitable rate constants exist for their design.

Headley and Davison (2003) has suggested that the exercise described in their paper could provide the starting point for the establishment of an Australian constructed wetland database leading to the further development and refinement of constructed wetland design models. This initiative is to be applauded and a concerted effort should be made to compile constructed wetland water quality data specific to the Western Australian climate and its native wetland species. Without this data constructed wetlands will continue to be designed using inappropriate models and their true potential remain unrealised.

References

Anda, M. (2005). Pers. comm., Mandurah, Timbers Edge development.

Crites, R. and Tchobanoglous, G. (1998). *Small and Decentralized Wastewater Management Systems*, McGraw-Hill, Singapore.

DLWC. (1998). *The Constructed Wetlands Manual*, Department of Land and Water Conservation, New South Wales, Australia.

Headley, T. R. and Davison, L. (2003). Design models for the removal of BOD and total nitrogen in reed beds. In: *On-site '03*, Armidale, NSW. pp. 169-176.

Kadlec, K. H. and Knight, R. L. (1996). *Treatment Wetlands*, CRC Press Inc. Boca Raton, Florida, USA.

Reed, S. C., Crites, R. W. and Middlebrooks, E. J. (1995). *Natural Systems for Waste Management and Treatment*, McGraw Hill.

Rousseau, D. P. L., Vanrolleghem, P. A. and De Pauw, N. (2004). Model-based design of horizontal subsurface flow constructed treatment wetlands: a review. *Water Research* 38, 1484-1493.

Russell, R. C. (1999). Constructed wetlands and mosquitoes: Health hazards and management options - An Australian perspective. *Ecological Engineering* 12, 107-124.

Turner, N., Heaton, K. and Meney, K. (1996). Use of innovative wastewater treatment methods and constructed wetlands to allow reuse of treated domestic wastewater. In: *Workshop on Wetland Systems for Wastewater Treatment*, Murdoch University. pp. 22-32.

USEPA. (1999). *Constructed Wetlands Treatment of Municipal Wastewaters Manual*. EPA.

Wong, T. H. F., Fletcher, T. D., Duncan, H. P. and Jenkins, G. A. (2006). Modelling urban stormwater treatment - A unified approach. *Ecological Engineering* 27, 58-70.

7 Case Studies

7.1 Hospital Wastewater Treated with Duckweed: A Case Study

M. Sirajul Islam[1*], M. Shahidul Kabir[2], Sirajul Islam Khan[3], M. Shafiqul Islam[1], Zahid Hayat Mahmud[1], Iqbal Kabir Jahid1, Sucharit Basu Neogi[1], Gopinath Balakrish Nair[1] and David Allen Sack[1].

[1]International Centre for Diarrhoeal Disease Research, Bangladesh (ICDDR,B), GPO Box 128, Dhaka-1000, Bangladesh.

[2]Department of Microbiology, Stamford University, 51, Siddeshory Road, Dhaka, Bangladesh.

[3]Department of Microbiology, University of Dhaka, Dhaka-1000, Bangladesh.

*Corresponding author:

Dr. Md. Sirajul Islam
Environmental Microbiology Laboratory
ICDDR,B, GPO Box 128
Dhaka-1000, Bangladesh
Phone: 880-2-8860523-32 Ext. 2407
Fax: 880-2-8823116, 880-2-8812529
Email: sislam@icddrb.org

Abstract

Duckweed (DW) is a tiny, fragile, free-floating aquatic plant which has been used for the treatment of wastewater and as fish feed. A study was carried out to determine the efficiency of duckweed in treating hospital based wastewater and reduction of chemical and microbial contamination of wastewater grown duckweed used as fish feed. The study was conducted at Kumudini Hospital Complex, Bangladesh. The abundance of vibrios and the level of faecal contamination of water, duckweed and fish from study and control ponds were monitored following standard procedures. A total of 343 samples including duckweed (107), water (132), fish gill (52) and fish intestinal content (52) were tested for vibrios and faecal coliform. Besides, every month stool samples collected from the handlers of duckweed and fish from both sewage and non-sewage areas were cultured for the members of Vibrionaceae and Enterobacteriaceae. The total dissolved solids (TDS) and pH were measured following the standard procedures. The results showed that the feacal coliform counts of raw sewage were 4.85 $\log_{10}$ CFU/ml which was reduced to 0.65 $\log_{10}$ CFU/ml after treating with duckweed. The TDS were also reduced from 483 mg/l in raw sewage to 222 mg/l after treatment. It was observed that duckweed can be used as a good means of wastewater treatment. The wastewater grown duckweed can be safely used as fish feed with minimum risk of transmission of pathogens to the handlers of the duckweed and fish. This study is therefore important from microbiological safety point of view of wastewater grown duckweed which is used as fish feed as well as the handlers who are involved in wastewater grown duckweed based fish culture project.

Introduction

Duckweed is tiny, fragile, free floating, aquatic plants. Duckweed belongs to four genera: *Lemna, Spirodella, Wolffia and Wolffiella*. About 40 species are known. None has distinct stems or leaves but consist of flattened, minute, leaf like, more or less oval "fronds", a few millimeters in diameter (Hillman, 1961; Vietmeyer, 1976).

Reproduction of duckweed is by vegetative means. The daughter fronds bud from reproductive pockets on the side of a mature frond. An individual frond may produce as many as 10 generations of progeny over a period of 10 days to several weeks before dying. Duckweed fronds can double their mass in two days. This is faster than any of the higher plants.

Duckweed has long been recognized for their potential as a source of high protein feed for animals. Studies conducted in various countries of the world e.g. USSR, USA, Canada and others have demonstrated the nutritional benefits of duckweed for both livestock and fish. Mirzapur duckweed project in Bangladesh has demonstrated that duckweed can be cultivated year round at rates exceeding 700 kg/ha/day. In Mirzapur duckweed project, one hectare of duckweed wastewater treatment plant yields up to one metric ton of fresh plants per day. This daily harvest can produce approximately 100 kg of fish or 100 kg of dried high protein duckweed meal.

Islam *et al.* (1990a) carried out a study on survival of toxigenic *Vibrio cholerae* O1 with a common duckweed, *Lemna minor*, in artificial aquatic ecosystems. They observed that toxigenic *V. cholerae* O1 can survive longer in association with duckweed (*L. minor*) than the water in which it was floating. Therefore duckweed which grows on the wastewater lagoons may harbour vibrios and other pathogenic bacterial flora, adsorbed from the water. When these plants are transferred from the lagoons to the ponds as fish feed, there is the possibility of transferring pathogenic bacteria along with the duckweed. Moreover, workers who are involved in this project may also contaminate themselves during handling of the duckweed and may contract diseases. The fish that eat the contaminated duckweed can also carry pathogenic bacteria on their surfaces and may also transmit the disease to fish handlers.

A project on *Lemnaceae* based wastewater treatment is going on at Mirzapur in Tangail where "Duckweed farming" is being done on agricultural land using either organic fertilizer or sewage collected from the Kumudini Hospital Complex. The hospital sewage is "treated" with duckweed by having it grown in a series of wastewater lagoons. Duckweed, when grown in these ponds, convert substantial amount of organic material and/or fertilizer into plant biomass; they convert nutrients and minerals dissolved in the water column into plant tissue.

When plants are harvested, nutrients and trace minerals are removed from the system and a dynamic nutrient and mineral sink is established. This forms the basis for a highly effective sewage treatment technology.

The duckweed which grow in sewage lagoons are harvested and used as the only source of food for fish. Therefore, duckweed is utilized for two purposes: one for treating wastewater and the other as food for growing fish. However, there are no data on the microbiological quality of both influent and treated effluent in Mirzapur duckweed project. There are also no data about the transmission of pathogenic organisms from the wastewater lagoons to fish ponds when the duckweed is transferred from the lagoons to the ponds where it is used as food for fish. Moreover, there are no data on health hazards involved among the workers who are directly involved with this project. Therefore, to answer all the above questions, this present study was undertaken.

Materials and Methods

Sampling locations and schedule

The sampling locations were situated within the Kumudini Hospital Campus of Mirzapur thana, under Tangail district of Bangladesh. The location is approximately 60 km north of Dhaka, the capital city of Bangladesh. In the first week of every month, duckweed (DW), water and fish samples were collected for a period of one year from the wastewater and control areas for the analysis of faecal coliform and isolation of vibrios. All the above mentioned sampling sites are shown in fig.1. From the wastewater area, DW and water samples were collected from 2 locations (L6 and L7) of wastewater lagoons and three fish growing ponds (P8, P9 and P10). Only water sample was collected from the raw sewage (L5). Tilapia (*Oreochromis nilotica*) was collected from fish pond no. P9 and Silver carp (*Hypothalamichthys molitrix*) was caught from fish pond no. P10 of wastewater area. From the control area, DW and water samples were collected from two DW growing beds (B1 and B2) and two of five fish growing ponds (CP3 and CP4). Tilapia and Grass-carp (*Ctenopharyngodon idella*) were caught from pond CP3 and CP4 respectively of the control area. The abbreviation B1 and B2 stand for duckweed growing beds in which artifical fertilizer are used as nutrients. The control pond numbers 3 and 4 are designated as CP3 and CP4. All these sampling sites are situated in the control area within the Hospital complex. In wastewater area, L5 stand for wastewater lagoon 5 from where the raw sewage is collected and no duckweed are grown. A total of 343 samples including duckweed (107), water (132), fish gill (52) and fish intestinal content (52) were tested for vibrios and faecal coliform.

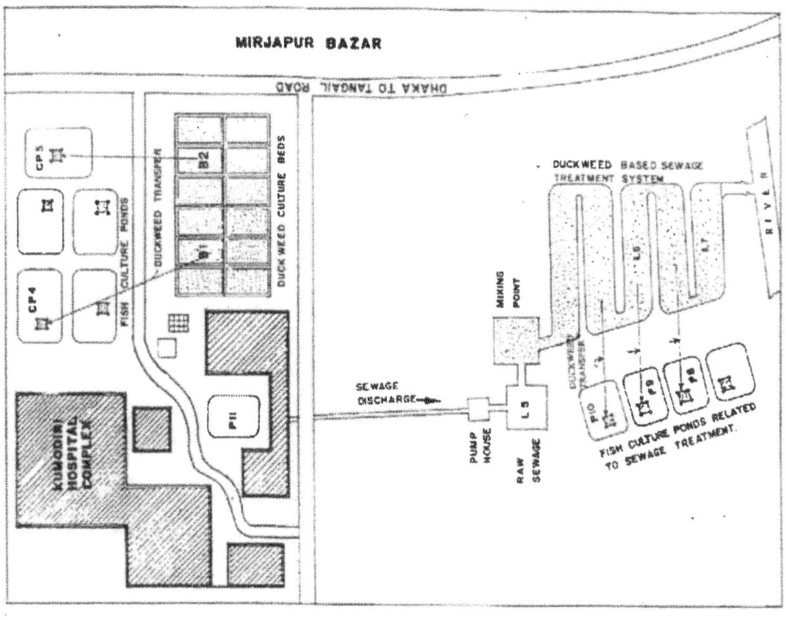

Figure 7.1.1: Schematic diagram of the sampling site at Mirzapure, Tangail, with specific sampling points. Sampling points are marked as B1, B2, CP3, CP4 in the control are and as L5, L6, L7, P8, P9 and P10 in the study area.

Measurement of chemical parameters

The total dissolved solids (TDS) and pH of water samples from different sampling sites were measured using portable meters (HACH, Model no. 41700/41800, USA and Orion Portable pH Meter, Cat. No. 210 A, Orion Research, USA) according to the instructions provided in user manuals.

Processing of samples

Water

A 0.1 ml water was taken directly from the collected sample or 10 fold dilutions were prepared and then 0.1 ml water was inoculated on mFC agar plate following drop plate technique (Hoben and Somasegaran 1982). When the counts were low, membrane filtration technique was followed to concentrate faecal coliforms. One ml water sample was passed through a membrane filter (0.22 μm pore size, 47 mm diameter, Millipore) and the filter with retained bacteria was placed on mFC agar plate which was incubated overnight at 44°C. Following incubation characteristic blue coloured colonies of faecal coliforms were counted and expressed as colony forming unit per milliliter (CFU/ml) of water sample following standard procedures (APHA, 1998).

A water sample of 50 ml was inoculated into 25 ml triple strength alkaline peptone water (3xAPW) for enrichment, incubated for 6 h and then plated onto thiosulphate citrate bile salt sucrose (TCBS) agar, taurocholate tellurite gelatin agar (TTGA) (Monsur, 1961) and inoculated plates were incubated at 37°C for 18 to 24 h. Then the members of Vibrionaceae were isolated and identified following procedures described earlier (Islam *et al.*, 1992a, Islam *et al.*, 1989).

Duckweed

Ten grams of DW was washed 3 times with normal saline for removing loosely attached bacteria. Then the washed duckweed was homogenized into a sterile electric blender for one minute in 90 ml normal saline. Ten fold dilutions of the homogenate in normal saline were prepared. From each dilution, 25 μL sample was plated directly on mFC agar following drop plate techique (Hoben and Somasegaran, 1982) and was incubated overnight at 44°C. Typical faecal coliform colonies were counted and expressed as colony forming unit per gram (CFU/g).

A 10 ml portion of the duckweed homogenates (1 g) was inoculated into 5 ml of triple strength alkaline peptone water (3xAPW) and identification of *V. cholerae* and *V. mimicus* was done following the procedures described earlier (Islam *et al.*, 1989).

Fish

Aseptically, fish gills were taken out and the ventral part of the fish was cut open with sterile scissors and the intestinal contents were collected. Ten g each of gills and intestinal contents were separately homogenized in 90 ml normal saline in electric blender for one minute. Homogenized gills and intestinal contents of fish were processed for fecal coliform, members of Vibrionaceae and Enterobacteriaceae following the procedures described above. Fishes were also regularly collected from the selected sites and carefully examined whether there is any ulcer due to infection.

Stool samples

Stool samples were collected from duckweed handlers every month during environmental sampling. Attempts were made to isolate common diarrhoeal pathogens of bacterial origin e.g., pathogenic vibrios, *Aeromonas* spp., *Plesiomonas shigelloides, Shigella* spp., *Salmonella* spp., and *Campylobacter* sp. using conventional culture technique following the procedures described earlier (Stoll *et al.*, 1982; Islam *et al.*, 1998, 1993a, 1993b, 1992b, 1991, 1990a).

Results

Figure 7.1.2 shows that the yearly average pH in raw sewage (RS) was 7.23 which increased to 7.40 and 7.68 in waste water lagoons No. L6 and L7 respectively. The average pH of sewage related fish pond (SRFP) was the highest (7.76) among all the waters tested in sewage area (study site). In non-sewage area (control site) the yearly average pH of water of duckweed bed (NSDB) and fish pond (NSFP) were 7.15 and 7.93 respectively. An increasing trend of pH from RS to SRFP was observed. Similarly the average pH in the non-sewage related fish pond (NSFP) was higher than the duckweed bed (DB).

Figure 7.1.3 shows that the yearly TDS in raw sewage was 483 mg/l which gradually decreased to 270, 240 and 222 mg/l in lagoons No. L6 and L7, and SRFP respectively in sewage area. In non-sewage related duckweed bed (NSDB) and NSFP the yearly average TDS was almost same.

Figure 7.1.4 shows that the faecal coliform counts in RS water was 4.85 $\log_{10}$ CFU/ml which reduced to 0.97 and 0.65 $\log_{10}$ CFU/ml in lagoons No. L6 and L7 respectively. The yearly average counts of faecal coliform in SRFP (1.3 $\log_{10}$CFU/ml), NSDB (1.53 $\log_{10}$ CFU/ml) and NSFP (1.23 $\log_{10}$ CFU/ml) were higher in comparison to lagoons No. L6 and L7.

Figure 7.1.5 shows that the yearly average faecal coliform counts gradually increased from 2.48 in sewage linked duckweed lagoon (SLDL) to 3.08, 3.44 and 4.14 $\log_{10}$ CFU/ml in SRFP duckweed, sewage related fish gill (SRFG) and sewage related fish intestine (SRFI) respectively. In non-sewage area, there was a similar increasing trend of faecal coliform counts, from 2.53 in non-sewage related duckweed bed duck weed (NSDB DW) to 2.75, 3.3 and 4.47 $\log_{10}$ CFU/ml in non-sewage related fish pond duckweed (NSFP DW), non-sewage related fish gill (NSFP FG) and non-sewage related fish intestine (NSFI) respectively.

Figure 7.1.6 shows that the yearly isolation rate of *V. cholerae* and *V. mimicus* in raw sewage (RS) were 83.3 and 8.3% respectively. In sewage linked duckweed lagoon (SLDL) water, the isolation rate of *V. cholerae* and *V. mimicus* were 29.2 and 12.5% respectively. The isolation rate of *V. cholerae* and *V. mimicus* in SLDL duckweed were 12.5 and 8.33% respectively. There was a substantial reduction of *V. cholerae* from RS (83.3%) to the SLDL water (29.2%).

In sewage related fish pond (SRFP), the yearly isolate rates of *V. cholerae* were 25, 28.6, 53.6 and 42.86% in water, duckweed (SRFP DW), fish gill (SRFP FG) and fish intestine (SRFP FI) respectively. The yearly isolation rate of *V. mimicus* were 2.77 and 2.86% in SRFP water and duckweed (DW) respectively. No *V. mimicus* could be isolated from fish gill but 7.14% *V. mimicus* was isolated from fish intestine (FI). In non-sewage duckweed bed (NSDB), the yearly average isolation rate of *V. mimicus* in duckweed and water was 4.16 and 8.33% respectively. No *V. cholerae* could be isolated from duckweed and water in NSDB. In non-sewage fish pond (NSFP), the yearly isolation rate of *V. cholerae* in water, duckweed, fish gill and fish intestine were 58.33, 50, 66.66 and 58.33% respectively. In NSFP, the isolation rate of *V. mimicus* in water, duckweed and fish gill was 8.33%, 4.16% and 12.5% respectively. No *V. mimicus* could be isolated from fish intestine in NSFP area.

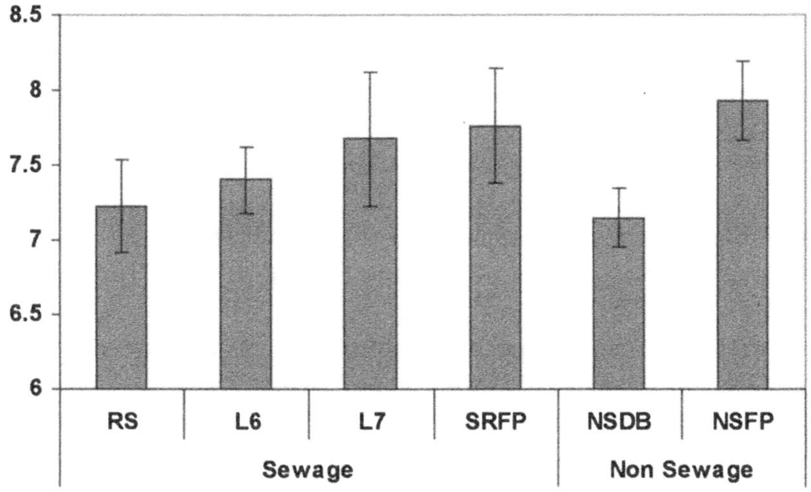

Figure 7.1.2: Average pH in different waters. RS= Raw sewage, L= Lagoon, SRFP= Sewage related fish pond, NSDB= Non-sewage related duckweed bed, NSFP= Non-sewage related fish pond. Each bar represents the arithmetic mean ±SD of 12 samplings.

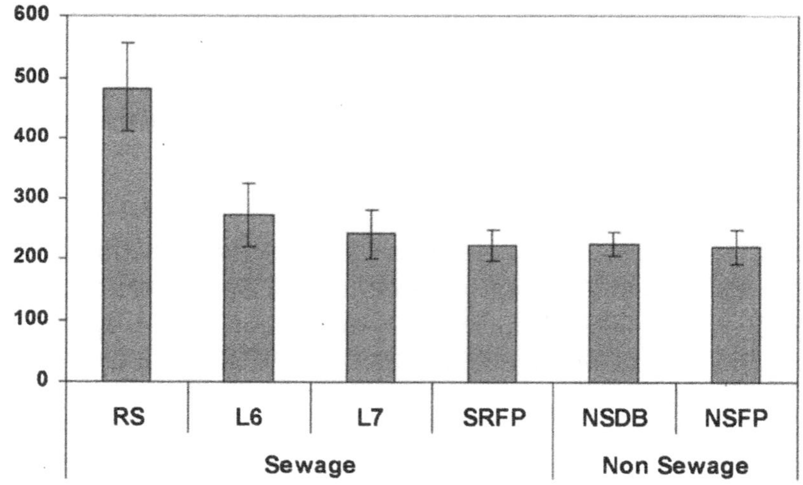

Figure 7.1.3: Average TDS of different waters. RS= Raw sewage, L= Lagoon, SRFP= Sewage related fish pond, NSDB= Non-sewage related duckweed bed, NSFP= Non-sewage related fish pond. Each bar represents the arithmetic mean ±SD of 12 samplings.

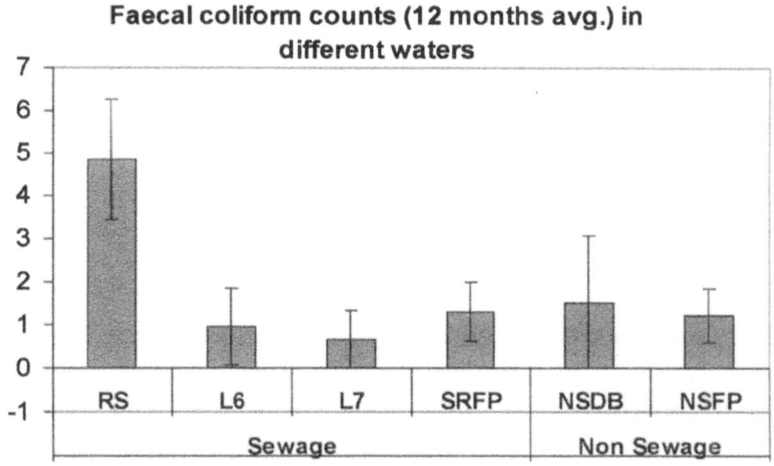

Figure 7.1.4: Faecal coliform counts (12 months avg.) in different waters. RS= Raw sewage, L= Lagoon, SRFP= Sewage related fish pond, NSDB= Non-sewage related duckweed bed, NSFP= Non-sewage related fish pond. Each bar represents the arithmetic mean ±SD of 12 samplings.

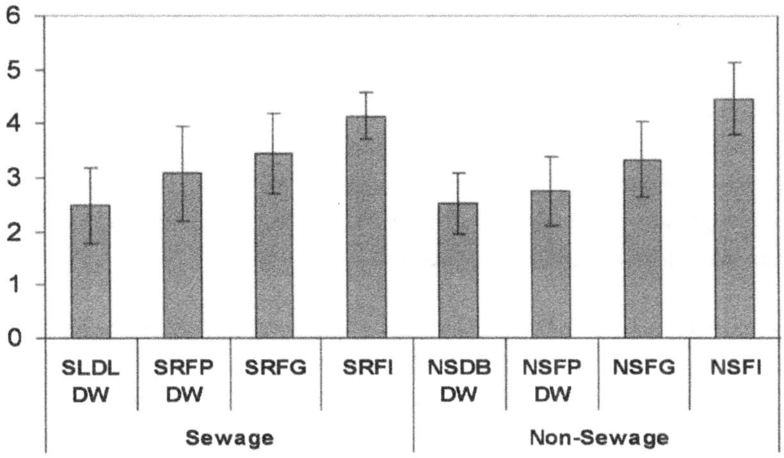

Figure 7.1.5: Faecal coliform counts (12 months Avg.) in different components. SLDL= Sewage linked duckweed lagoon, SRFP= Sewage related fish pond, SRFG= Sewage related fish gill, SRFI= Sewage related fish intestine, DW= Duck weed, NSDB=Non-sewage related duckweed bed, NSFP= Non-sewage related fish pond, NSFG= Non-sewage related fish gill, NS FI= Non-sewage related fish intestine. Each bar represents the arithmetic mean ±SD of 12 samplings.

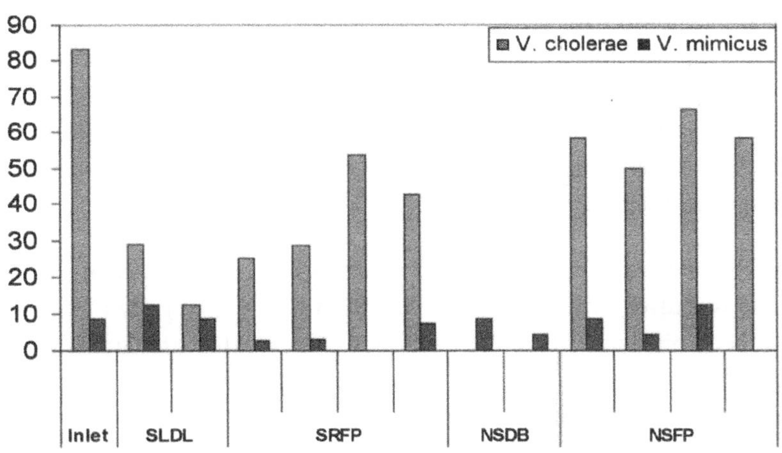

Figure 7.1.6: Vibrio abundance among different components. RS= Raw sewage, SLDL= Sewage linked duckweed lagoon, SRFP= Sewage related fish pond, FG= Fish gill, FI= Fish intestine, DW= Duck weed, NSDB=Non-sewage related duckweed bed, NSFP= Non-sewage related fish pond.

Discussion

This study demonstrated that duckweed can efficiently reduce the faecal coliform counts in wastewater (Fig. 4). The results also showed that there was no significant difference in faecal coliform counts in water between sewage and non-sewage fish ponds (control ponds). The faecal coliform counts in duckweed, fish gill and fish intestine were also similar in both sewage and non-sewage related ponds (Fig. 5).

Though the fish ponds in both sewage and non-sewage areas harbour *V. cholerae,* no handler in both areas had ever contracted any infection with *V. cholerae*. There occurrence of *V. cholerae* in both sewage and non-sewage related areas (Fig. 6) indicates that the presence of this bacterium is not related with sewage connection. In fact, *V. cholerae* are the normal flora in the aquatic environment of Bangladesh (Islam *et al.*, 1994a, 1992a). Not only *V. cholerae* but *V. mimicus* was also abundant in both wastewater area and non wastewater area, but no infection has ever been found among the handlers with *V. mimicus* as well. Similarly the faecal coliform counts in DW, FG and FI were always higher in fish ponds than the wastewater lagoons and non-sewage fish pond and DB (Fig. 4 and 5). However, no fish had ever been found to be infected with any pathogen.

Therefore, this study suggests that duckweed can be grown in sewage lagoons and can be safely used as fish feed with minimal risk for transmission of enteric pathogens from the fish. The workers who handle both the contaminated duckweed grown in sewage lagoons and the fish fed by sewage grown duckweed were also free from diarrhoeal pathogens. Therefore, the sewage grown duckweed and fish fed on these duckweed did not pose any health hazard to the handlers as well as to the fish.

The results also demonstrated that hospital based waste water can be treated chemically and microbiologically using duckweed. There is a substantial reduction of TDS from raw sewage (RS) to lagoons L6 and L7 (Fig. 3). In non-sewage related duckweed bed and NSFP, there was no difference in TDS. The pH was higher in sewage lagoons and SRFP as well as in non-sewage related FP than RS (Fig. 2). The higher pH might help *V. cholerae* to multiply in fish ponds than in water and duckweed of wastewater lagoons as vibrios are halophilic organism (Islam *et al.,* 1990c). As a result, the isolation rate for *V. cholerae* was higher in water and duckweed in SRFP (Fig. 6). The isolation rate was also higher in fish gill (FG) and fish intestine (FI) than water and DW. The high proportion of *V. cholerae* in fish intestine might be related to their consumption of different aquatic flora and fauna (e.g., phytoplankton, zooplankton, duckweed etc.) which can harbour this organism (Islam *et al.*, 1994b, 1992a, 1992b, 1990b). Production of mucinase by *V. cholerae* might be responsible for its attachment to mucilaginous plant cells (Islam *et al.*, 2002). As, fish gills generally contain high number of plankton (because of filtering huge volume of water for respiration) to which *V. cholerae* may remain attached, that might explain higher rate of isolation of *V. cholerae* from fish gill.

The present study also showed that duckweed can be grown in wastewater lagoons which will help to treat the wastewater at the same time the wastewater grown duckweed can be used safely as fish feed for pisciculture. Therefore, use of wastewater grown duckweed as fish feed is more profitable than using duckweed grown by artificial fertilizer. This study provided some vital information about microbiological safety and economic benefit of using sewage for growing duckweed which can be used for pisciculture.

We can conclude from this study that in developing countries, duckweed can be used as a good means for sewage treatment. The duckweed can also be used as fish feed. This study is therefore important from the microbiological safety point of view of sewage grown duckweed which is used as fish feed as well as the handlers who are involved in such duckweed based fish culture project.

Acknowledgement

This research was funded by the Swiss Development Co-operation (SDC). ICDDR,B acknowledges with gratitude the commitment of SDC to the Centre's research efforts. The authors are grateful to Mr. Md. Ekramullah and the authority of Kumudini Hospital, Mirzapur for supporting this study.

References

APHA, AWWA and WEF. (1998). Microbiological Examination. *In* Standard methods for the examination of water and wastewater. 20th edition. *Edited by* A.E. Greenberg, L.S. Clesceri and A.D. Eaton. American Public Health Association, American Water Works Association, Water Environment Federation, Washington D.C., U.S.A.

Hillman, W.S. (1961). The Lemnaceae, or Duckweed. A review of the descriptive and experimental literature. Botanical Rev. 27: 221-287.

Hoben, H.J. and Somasegaran, P. (1982). Comparison of the pour, spread, and drop plate methods for enumeration of Rhizobium spp. in inoculants made from presterilized peat. Appl. Environ. Microbiol. 44: 1246-1247.

Islam, M.S., Goldar, M.M., Morshed, M.G., Khan, M.N.H., Islam, M.R. and Sack, R.B. (2002). Involvement of the *hap* gene (mucinase) in the survival of *Vibrio cholerae* O1 in association with the blue-green alga, *Anabaena* sp. Canadian Journal of Microbiology. 48: 793-800.

Islam, M.S., Hossain, M.S., Hasan, M.K., Rahman, M.M., Fuchs, G., Mahalanabis, D., Baqui, A.H. and Albert, M.J. (1998). Detection of shigellae from stools of dysentery patients by culture and polymerase chain reaction techniques. Journal of Diarrhoeal Disease Research 16 (4): 248-251.

Islam, M.S., Drasar, B.S. and Sack, R.B. (1994a). The aquatic flora and fauna as reservoirs of *Vibrio cholerae* : A review. Journal of Diarrhoeal Diseases Research 12(2): 87-96.

Islam, M.S., Miah, M.A., Hasan, M.K., Sack, R.B. and Albert, M.J. (1994b) Detection of non culturable *Vibrio cholerae* O1 associated with a cyanobacterium from an aquatic environment in Bangladesh. Trans Roy Soc Trop Med Hyg. 88: 298-299.

Islam, M.S., Hasan, M.K. and Khan, S.I. (1993a). Growth and survival of *Shigella flexneri* in common Bangladeshi foods under various conditions of time and temperature. Applied and Environmental Microbiology 59: 652-654.

Islam, M.S., Hasan, M.K., Miah, M.A., Qadri, F., Yunus, M., Sack, R.B. and Albert, M.J. (1993b). Isolation of *Vibrio cholerae* 0139 Bengal from water in Bangladesh. Lancet. 342: 430.

Islam, M.S., Alam, M.J. and Neogi, P.K.B. (1992a). Seasonality and toxigenicity of *Vibrio cholerae* non-O1 isolated from different components of pond ecosystems of Dhaka city, Bangladesh. World Journal of Microbiology and Biotechnology 8(3):160-163.

Islam, M.S., Alam, M.J. and Tzipori, S. (1992b). Abundance of *Aeromonas* spp. in various components of pond ecosystems in Dhaka, Bangladesh. International Journal of Environmental Studies 39: 297-304.

Islam, M.S., Alam, M.J. and Khan, S.I. (1991). Distribution of *Plesiomonas shigelloides* in various components of pond ecosystems in Dhaka, Bangladesh. Microbiology and Immunology 35(11): 927-932.

Islam, M.S., Drasar, B.S. and Bradley, D.J. (1990a). Survival of toxigenic *Vibrio cholerae* O1 with common duckweed, *Lemna minor,* in artificial aquatic ecosystems. Tran Roy Soc Trop Med Hyg. 84: 422 424.

Islam, M.S., Drasar, B.S. and Bradley, D.J. (1990b) Long- term persistence of toxigenic *Vibrio cholerae* O1 in the mucilaginous sheath of a blue green alga, *Anabaena variabilis.* Journal of Tropical Medicine and Hygiene 93: 133-139.

Islam, M.S. (1990c). Effect of various biophysicochemical conditions on toxigenicity of *Vibrio cholerae* O1 during survival with a green alga, *Rhizoclonium fontanum* in artificial aquatic environment. Canadian Journal of Microbiology 36:464-468.

Islam, M.S., Alam, M.J., Siddique, A.B. and Khan, S.I. (1989). Abundance of members of vibrionaceae in various components of ponds in Dhaka City, Bangladesh. Bangladesh journal of Microbiology 6(1): 52-3.

Monsur, K.A. (1961). A highly selective gelatin taurocholate tellurite medium for isolation of *Vibrio cholerae*. Tran Roy Soc Trop Med Hyg. 55:440 445.

Stoll, B. J., Glass, R.I., Huq, M.I., Khan, M.U., Holt, J.E. and Banu, H. (1982). Surveillance of patients attending a diarrhoeal disease hospital in Bangladesh. Br. Med. J. 285: 1185-1188.

Vietmeyer, N.D. 1976. Making aquatic weeds useful: some perspectives for developing countries. National Academy of Science. Washington, D.C., U.S.A. pp. 127-129.

8 Rain Water Systems

8.1 An Analysis of the Operational Data of Rainwater Utilization Facility in a Dormitory in Seoul National University

Sungwon Park[1], Sangrae Kim[1], Haryoung Kim[1], and Mooyoung Han[1]

[1]The school of Civil, Urban & Geosystems Engineering, Seoul National University, San 56-1, Shillim-dong, Gwanak-gu, Seoul, 151-742, Korea,
(E-mail: myhan@snu.ac.kr, gocine@snu.ac.kr, swpark@waterfirst.snu.ac.kr, bapresi@naver.com)

Abstract

When rainwater utilization facility is installed and operated, the most important thing to be considered is the estimated method of profit capacity of storage tank. It's determined by operation factors of rainwater utilization facility: rainwater utilization ratio, ratio substitution of tapwater, system cycling and the day of rainwater utilization.

In this research, applicability and efficiency of rainwater system operation factor diagram are examined by comparing with real operational data in SNU dormitory for two years. And expected operation problems are suggested in order to help installation and operation of rainwater utilization facility in newly constructed and extended building.

Key Words

Rainwater harvesting, Operational data, Operation factors, Rainwater storage tank

Introduction

Calculated method of profit capacity of rainwater storage tank is one of the most important things in rainwater utilization systems. Because installation and operation cost is actually decided as the storage tank volume (Han *et al*, 2004). Making decision of storage tank is affected by basic factors such as rainfall depth, catchment area and rainwater usage. It must be decided on operation factors (rainwater utilization ratio, ratio substitution of tapwater, system cycling and the day of rainwater utilization) of rainwater utilization facility, which are considered as basic factors.

At present, the study deciding profit rainwater storage tank volume is in the being stage. So the theoretical and scientific data is deficient. Considering that the importance of storage tank volume, empirical study for deciding tank volume is necessary.

So in this study, operational data in SNU dormitory is analyzed for two years, 1) for usefulness of rainwater operation factor diagram, comparative analysis is examined by cross checked real and simulation data. 2) By suggesting expected operation problems, installation and operation of rainwater utilization facility, newly constructed and extended building rainwater facilities can be facilitated.

Materials and Methods

Status of Research Sites

The 1st rainwater system in Korea has been operated for toilet flushing water and gardening water at a dorm of Seoul National University (SNU) built in November, 2003. Graywater was supposed to be supplied to toilet based on the design concept of the dormitory, but it was changed into a rainwater system during construction. Therefore, two of four buildings are being used by rainwater and the others are being used by graywater **(Fig. 1)**.

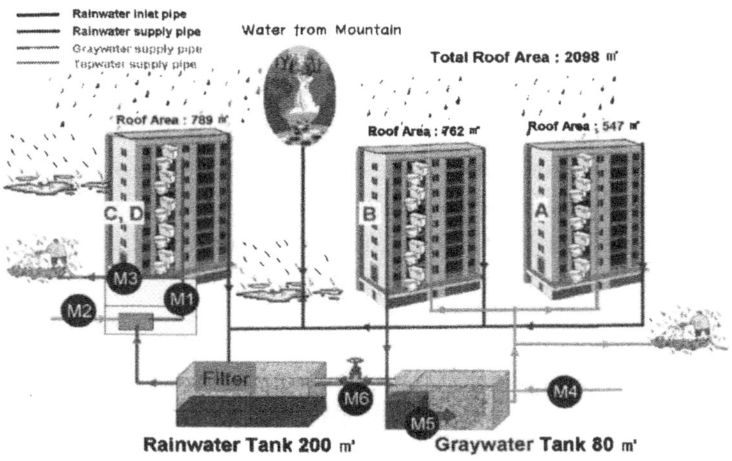

Figure 8.1.1: Schematic diagram of rainwater utilization facility in a dormitory in SNU

The rainwater system started operating on April 27th 2004. The purposes and benefits of rainwater system in dorm in SNU are therefore:

- runoff control, inundation prevention;
- reduction of tapwater usage, insurance against water for emergency;
- environmental education, research;

Besides, it is significant for the 1st working facility to use rainwater as flushing water, in connection with graywater system and valley, and be a model rainwater system for future.

Measuring method of water quantity in a dormitory

Water quantity in rainwater system at SNU dorm is monitored for 2 year from April 2004 to December 2005. Water quantities of rainwater, tapwater, and graywater are measured by remote monitoring system everyday.

Diagram of rainwater operation factors (Han *et al*, 2004).

Rainwater operation factors were determined according to the design of rainwater utilization system. Factors that affect rainwater operation are catchment area, storage volume, rainwater usage, and etc. Fig. 4 shows the effect operation factors and storage volume/catchment area, rainwater usage per day/catchment. In this study, simulation carried out in the range of 0.001 ≤ (Rainwater usage per day /Catchment area) ≤ 0.05, 0.002 ≤ (Rainwater storage volume / Catchment area) ≤ 0.2. This means that efficiency can be founded in diagram as the range of water usage per day from 0.1 ton to 5 ton and the range of storage tank volume from 0.2 ton to 30 ton when the catchment area is 100m^2. This graph can be applied to decide tank volume for average rainfall base on 5 years rainfall data in Seoul.

Result and Discussion

Comparative analysis between operational data and simulation data

Operational data of rainwater utilization facility in a dormitory

Rainwater is supplied or toilet flushing in 2 buildings at a dorm in SNU for 2 years from April 2004 to December 2005. As shown in **Fig.8.1.2**, total amount of rainwater usage was 2,628 tones, and the rainwater system cycled 13 times. Total usage of tapwater and rainwater increased rapidly in summer. It shows characteristics (monsoon, concentration of rainfall in summer) of climate in Korea. Fig. 3 shows that monthly usage of rainwater and tapwater in detail and that the portion of rainwater in summer is very high but that of tap water in winter is very high. **Table.8.1.1** shows the operational data of rainwater facility in dormitory.

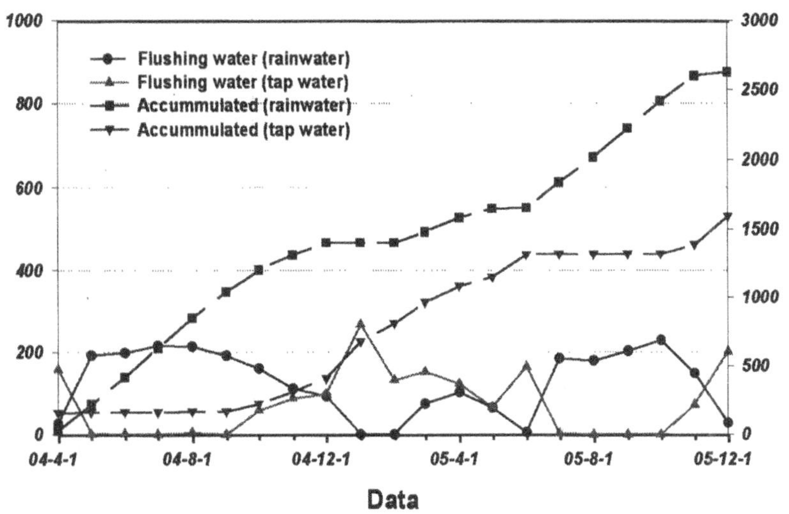

Figure 8.1.2: Changes in monthly water quantity of used and accumulated rainwater at the rainwater system

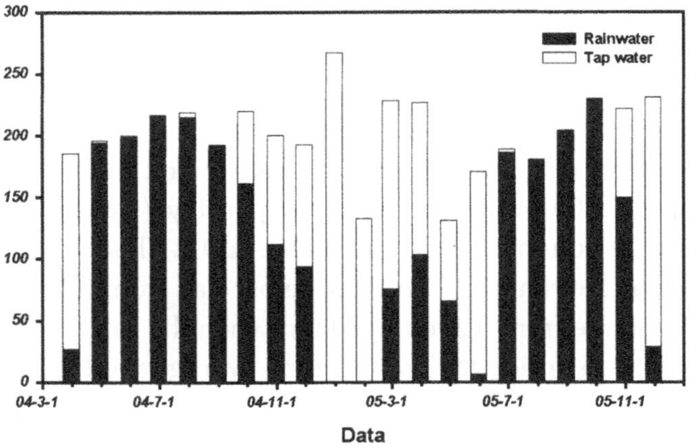

Figure 8.1.3: Changes in monthly water quantity of used rainwater and tapwater at the rainwater system

Table 8.1.1: Operation data of rainwater utilization facility in a dormitory

	Rainfall	Collected rainfall	Rainwater usage	Domestic water usage
2004	1,499m^2	2,998m^2	1,405m^2	20,946m^2
2005	1,358m^2	2,716m^2	1,223m^2	25,817m^2
Total	2,857m^2	5,714m^2	2,628m^2	46,763m^2

Simulation data of rainwater utilization facility in a dormitory

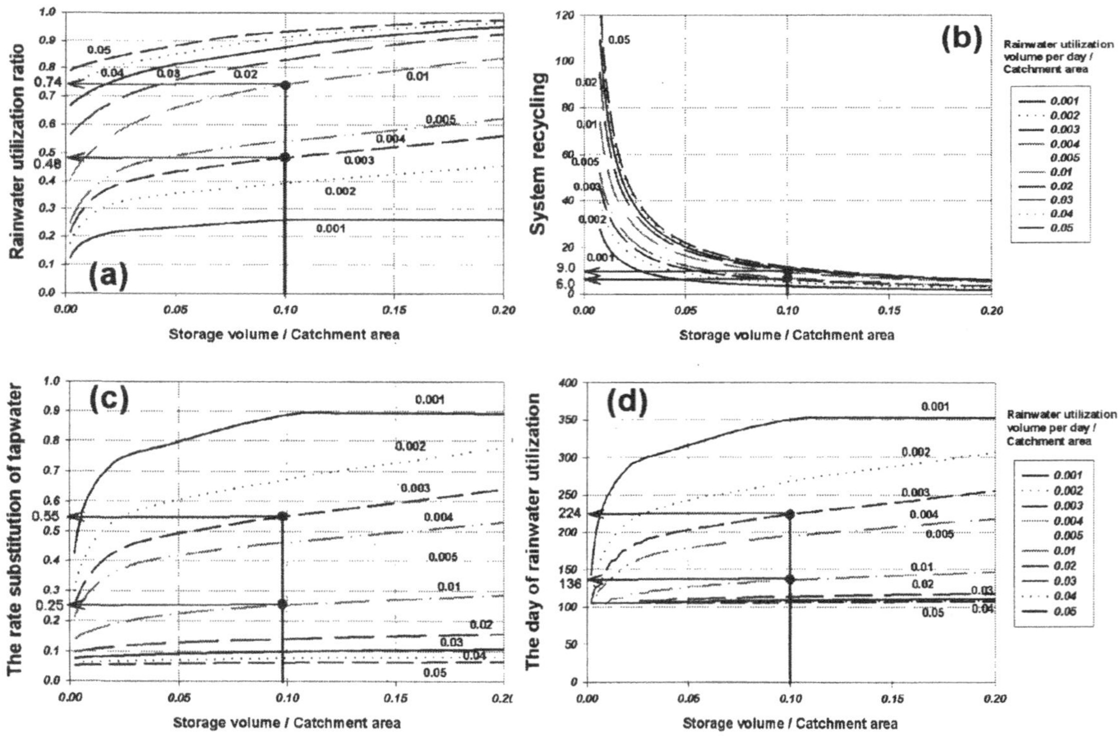

Figure 8.1.4: Rainwater operation factors (a) rainwater utilization ratio, (b) system recycling, (c)the rate of substitution of tapwater, and (d) the day of rainwater utilization.

In Fig. 4 following conditions are applied to rainwater utilization facility in SNU dormitory from Rainwater operation factors diagram.

- Catchment area: 2,098m^2
- Rainwater usage per day: 6.2m^2/day (B/D C, D), 22m^2/day (B/D A, B, C, D)
- Rainwater storage volume: 200m^2
- Rainwater usage / Catchment area = (6.2m^2/day)/(2,098m^2) = 0.003 m/day, (22m^2/day)/(2,098m^2) = 0.01 m/day
- Rainwater storage tank volume /Catchment area = (200m^2)/(2,098 m^2) = 0.1 m

From above conditions 4 rainwater operation factors are applied.

Comparative analysis between operational data and simulation data

Rainwater operation factors can be obtained by using real operational and simulation data. Table 8.1.2 shows rainwater operation factors.

Equation induces each rainwater operation factor is following.

1. Rainwater utilization ratio (%) = $\frac{\text{Rainwater usage per year}}{\text{Collected rainfall per year}} \times 100$

2. The ratio substitution tapwater (%) = $\frac{\text{Rainwater usage per year}}{\text{Domestic water usage per year}} \times 100$

3. System recycling = $\frac{\text{Collected rainfall per year}}{\text{Rainwater storage volume}}$

4. The days of rainwater utilization = The number of days remained in rainwater storage tank (The number of available rainwater)

Table 8.1.2: Rainwater operation factors of case study and simulation

	Rainwater utilization ratio	The ratio substitution of tapwater	System recycling	The days of rainwater utilization
2004	47%	6.7%	7	214
2005	45%	4.7%	6	198
Simulation-1	48%	55%	6	224

Table 8.1.3: Rainwater operation factors of simulation for a differential rainwater usage per day

	Rainwater utilization ratio	The ratio substitution of tapwater	System recycling	The days of rainwater utilization
Simulation-1	48%	55%	6	224
Simulation-2	74%	25%	9	136

To comparatively analyze annual data, real operational data is separated into two parts, for year 2004 and 2005. In simulation-1 rainwater usage is 6.2m^2/day (supply B/D C, D: now operated) and in simulation-2 it is 22m^2/day (supply B/D A, B, C, D: assumption). All the parameters for year 2004 have higher values. Although annual rainwater usage is nearly similar, but rainfall depth of 2004 is 140mm higher than 2005.

In table 8.1.2 rainwater utilization ratio, system cycling and days using rainwater but the ratio substitution of tapwater of real operational data is nearly similar with that of simulation. Rainwater utilization ratio is about 50% and it means the amount of rainwater flowed out to sewerage is about 50%. That rainwater utilization ratio is higher, means it is possible to collect more rainwater that falls on the catchment area.

Real data and simulation is very different in the ratio substitution of tapwater because of the characteristic of dormitory. In case of changeable total usage, larger the storage tank volume is, the higher the ratio substitution of tapwater is. But in dormitory comparatively many people live in small area. So in dormitory total water usage is much higher than rainwater usage. When total usage increases, rainwater usage also increases. But increasing ratio of rainwater usage is much smaller is comparison to the ratio substitution of tapwater decrease. Likewise when total water usage decreases, the ratio substitution of tapwater increases much higher.

In table 8.1.3, changing rainwater usage is simulated. As rainwater usage increases, rainwater utilization ratio and system cycling increases by, but the ratio substitution of tapwater and days using rainwater decreases by half. In operating rainwater utilization facility, it's important to make the ratio substitution of tapwater higher, because it means that dependency is low and independency of building is high.

In the case of dormitory, making larger volume tank to increase the ratio substitution of tapwater is not effective as Fig.4 (c) shows. It's inefficient in cost. In this study, to increase the ratio substitution of tapwater, decreasing total water usage is more much more important than making larger volume tank.

As foresaid diagram shows wide-range to lower total water usage. But the ratio substitution of tapwater graph must be modified to decide the profit tank volume applying diagram for high total water usage. So modified diagram applying to high total water usage should be suggested later.

Design and operational suggestion

Installations

For the installation of rainwater system, a various technical knowledge should be put together into the design concept. For example, architecture for designing rooftop and drain, civil engineering for establishing rainfall intensity and ground condition, sanitory engineering for installing pump, pipe, or filter, and landscape for designing a pond should be considered for the system.

In case of this rainwater system, having considered the location and ground condition enough at the planning concept would have saved the cost for the construction and management. Because of low overflow level, rocky ground condition and distance from nearby sewerage caused an additional cost for installation and operation.

Besides, it was necessary to instance standard to install the rainwater system. In this case, it should cost too much for the construction by applying useless or unsuitable pipe.

Management

The rooftop as catchment area should be maintained by cleaning regularly because it could influence on the water quality of storage tank. In this system, there were several problems to maintain clean rooftop and were several ways to solve.

First of all, the entrance and exit of students to the rooftop contribute pollution. The most of garbage by students could be filtered but some pollution passed through the filter. So these inflow contaminant could affect quality of the rainwater in the storage tank. In fact, the water quality was improved after keeping students from entering to the rooftop on the gate by putting "No entrance" sign.

In addition, there was an antenna on the rooftop. Birds used it for their resting place. So birds' excretion with rainwater introduced into the storage tank and caused infection of bacteria to the stored rainwater (B.W Ritchie *et al*, 1994). Therefore, It is intended the shape of antenna, or to divert the rainwater around the antenna to enter in catchment pipe for the rainwater

Conclusion

Diagram is made to solve the rainwater operational factors (rainwater utilization ratio, the ratio substitution of tapwater, system cycling and days of using rainwater) to apply for installation and operation of rainwater utilization facility in Seoul. On the basis of this study, applicability and efficiency of rainwater system operation factor diagram are examined by comparing with real operational data in SNU dormitory for two years. Finally expected operation problems are suggested in order to help installation and operation of rainwater utilization facility in newly constructed and extended building.

1) Comparative analysis between real data and simulation: all items but the ratio substitution of tapwater of 2004 and 2005 is nearly similar. It's because in dormitory, total water usage is comparatively much higher than rainwater usage. To increase the ratio substitution of tapwater, it's the best method to decrease total water usage. Rather making larger volume because of cost inefficient. The diagram is well applied to low water usage. But to apply diagram to high water usage it should be modified. So modified diagram applying to high total water usage should be suggested later.
2) Design and Operational suggestion: Suggested from installation and operation of rainwater system, civil engineering, architecture, facility engineering and landscape should be put together into the design concept, and the guideline for installation is necessary. Besides, rooftop as catchment area should be maintained because it could influence on the water quality of storage tank.

Therefore, these results will be useful for people who apply and operate a rainwater system.

Reference

B.W Ritchie, G.J.Harrison, L.R.Harrison (1994), AVIAN MEDICINE: PRINCIPLES AND APPLICATION, Wingers Publishing, Inc., pp950-951.

Han, Mooyoung (2004), Rainwater Harvesting: A new paradigm to meet MDG and sustainability, Rainwater Harvesting Workshop in the Congress IWA, pp1-8.

Han, Mooyoung, Han, Myoung-sil, Kim, Sangrae (2004). A Consideration in Determining the Tank Volume of Rainwater Harvesting System in Building, J. of Korean Society of water and Wastewater, 18(2), pp. 99-109.

Han, Mooyoung, Youngwan Kim, Sangrea Kim (2003), Rainwater system in dormitory at Seoul National University (in Korean), the 3rd international workshop on rainwater harvesting, UNEP-SNU Rainwater research center, pp.105-112.

Kim Youngwan (2005) The Design plans considering particle characteristics and behavior in rainwater storage tank (in Korean). Master's thesis, the school of civil, urban & geosystems engineering, Seoul national university.

Ministry of Environment (2002), statistics for water service (in Korean).

9 Composting Toilets

9.1 Mass Balance In the Treatment of Blackwater with Mechanical Composting Toilet by Using Sawdust as a Matrix

Shinya Hotta, Ryusei Ito, Naoyuki Funamizu

Department of Environmental Engineering, Graduate School of Engineering, Hokkaido University, Kita-13, Nishi-8, Kita-ku, Sapporo 060-8628 Japan

Tel.: +81 117066272; fax: +81 117066272.

zhenye@eng.hokudai.ac.jp; ryuusei@eng.hokudai.ac.jp; funamizu@eng.hokudai.ac.jp;

Abstract

Mechanical composting toilet by using sawdust as a matrix had been performed for 50 days. Bioreactor in the mechanical composting toilet contained the mixing and heating device to promote composting reaction of feces and urine (toilet waste). Furthermore ventilation system was also equipped with the bioreactor that induces airflow from the toilet seat to outside. Daily amount of human feces (150g/day) and urine (1000ml/day) per person was fed to the sawdust matrix. Mass balance of water, carbon and nitrogen was obtained in the operation. Power consumption in two points (heating device only and total system) has also been monitored. The operation was continued without replacement of the sawdust matrix during the operation for 50days. During the operation, 54kg of input water mainly from urine was removed from the matrix because of water evaporation rate accelerated by heating. Accumulation rate of carbon and nitrogen in the matrix were obtained by 8.75g-C/day and 1.82g-N/day respectively. Approximately 55% of carbon and 25% of nitrogen in input remained in the matrix for 50days operation.

Introduction

In domestic wastewater, toilet waste contains the very unique characteristic. Toilet waste contains a huge amount of nutrient, organic matter (Almedia *et al.*, 1999). An efficient composting toilet is therefore a key piece in the technology of the onsite wastewater differentiable treatment system (OWDTS) (Lopez, *et al.*, 2002a). The composting toilet by using sawdust as a matrix is a dry closet for toilet waste that requires no water for flush and transportation of toilet waste. The composting toilet called bio-toilet contains several mechanical devices to accelerate the composting process of toilet waste. Temperature and frequency of mixing is automatically controlled. The bio-toilets have already been used in several public areas without sewage pipe in Hokkaido island Japan. Kitsui and Terazawa reported successful operation of the bio-toilet in Japan. They discussed contribution of the bio-toilet to a nutrient recycle of N, P and K required in farmland (2001). However, it is well known that huge amount of nitrogen is lost in the composting process of livestock manure by the volatilization of ammonia (Martins and Dews, 1992). There are several differences in operational conditions between conventional composting of live stock manure and composting of toilet waste in the bio-toilet. For instance, the bio-toilet receives toilet waste almost everyday with lower load of organic matter in the matrix.

Reaction temperature and frequency of mixing are well controlled automatically in the bio-toilet. Lopez *et al.*, have reported the aerobic biodegradation of feces by using sawdust as a matrix (Lopez *et al.*, 2002b) and they have developed mathematical model to describe the composting process of human feces (Lopez *et al.*, 2004a, 2004b). Hotta *et al.*, have reported the nitrogen transformation during the aerobic biodegradation of feces (Hotta *et al.*, 2005). In their reports, conditions for the composting process were strictly controlled, for instance, around 60% of moisture content, 100ml/min of aeration, 50oC of reaction temperature and relatively high load of organic matter against the sawdust matrix. However it is impossible to prepare such a perfect control of the condition in the bioreactor under practical operation. Therefore it is easily assumed that composting process, at least composting rate, may be unique in the practical operation. The bio-toilet consumes lots of power in the devices of heating, mixing and ventilation system to maintain the suitable condition for composting process. So far, there is report concerning about power consumption in the practical operation of the bio-toilet. Furthermore composting process under practical condition has never been reported well in literature. Verification by using the real composting toilet is therefore required to apply the knowledge and experience in previous reports to the practical situation. In previous studies for the bio-toilet (Lopez *et al.*, 2005), design of the bioreactor of the bio-toilet was mainly decided by water and carbon in the sawdust matrix. Nitrogen transformation of ammonification is also interesting research topic as regards to nutrient recovery. The aim of this study is therefore to obtain the mass balance of water, carbon and nitrogen with continuous feeding of toilet waste to the sawdust matrix under practical operation. Further consumption of power is also measured for the evaluation of the bio-toilet under operation.

Materials and methods

Materials

Virgin sawdust from a coniferous tree was prepared as a matrix. It had been dried at room temperature for a couple of days; initial moisture content was 7.5%. Then it was passed through a sieve of 2.8mm to remove a large size particle. Human feces were collected from young healthy males and the feces for a week was mixed in a plastic bucket and stored at refrigerator at 5°C. As regards to fecal characteristic, there were no variations on the contents of moisture, nitrogen and carbon: their average values were 83%, 69mg-N/g-dry and 491mg-C/g-dry respectively. Human urine was also prepared in the same way for human feces. Concentrations of urinal carbon and nitrogen in daily input were 10.2g-N/l and 4.0g-C/l respectively. Since previous data had given approximately 95% of the moisture content of urine (Environmental Technology Association, 1981), we applied it to the calculation in this study. Since several reports have shown daily amount of human feces and urine ranges 100-150g-wet and 1000-1500ml respectively (Environmental Technology Association, 1981), 150g-wet of feces and 1000ml of urine were fed during the operation. Table9.1.1 lists the daily input of the toilet waste to the bio-toilet: total input, for 50days operation, of carbon (805g), nitrogen (595g), mass (3.7kg) and water (53.8kg) were calculated.

Table 9.1.1: Daily inputs to the sawdust matrix

	Feces	Urine	Total
Nitrogen	1.7	10.2	12
Carbon	12.1	4.0	16
Mass	24.7	50.0	75
Water	125	95.0	1075

The bio- toilet

A composting toilet for one person was installed inside the laboratory. Several devices were equipped with the toilet to conduct the heating and mixing of the sawdust matrix. Further one way of airflow was also produced by ventilation system. Since heating device was on the bottom of the composting reactor in the toilet, temperature of the matrix was not uniformed even after mixing. It usually ranges from 30 to 60°C under normal condition. Mixing frequency was set up to once an hour. Power consumptions for heating device only and whole were separately measured.

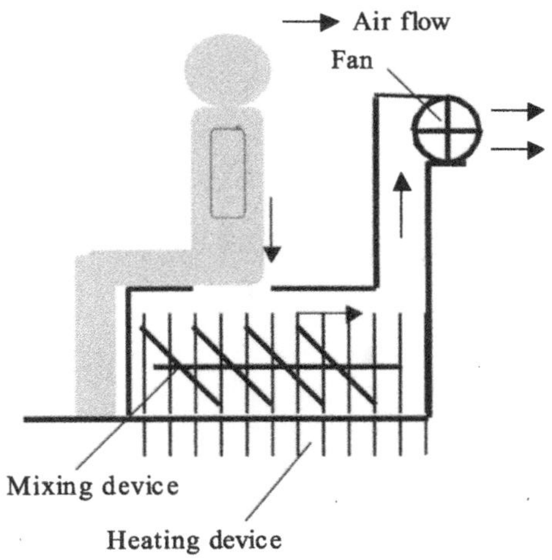

Figure 9.1.1: Schematic of the bio-toilet

Analysis

Every 10days during the operation, sawdust matrix was taken from the composting reactor and weighed, then 35g-wet of sawdust was obtained as sample. Total loss by the sampling was approximately 175g-wet. Since initial amount of sawdust matrix was approximately 3700g-dry, it therefore did not significantly affect an evaluation of mass balance in the matrix. Moisture content of the sawdust matrix and feces was measured by weight difference between a wet-sample and dried one that was obtained by using oven at 105°C for 24hours. Dried sample was used for elemental analysis (SHIMAZU: SUMIGRAPH NC-1000) to obtain contents of carbon and nitrogen in solid. Total ammonia and Kjeldhal nitrogen was measured by using an extraction. The extraction was obtained by filtration of mixture with 5g-wet sample and 50ml of distilled water. In a process of the filtration, we added more distilled water to extract completely from the sawdust matrix until total volume of the extraction reached 1000ml. We followed the indophenol method for determination of ammonia (Kempers *et al.*, 1986) and semi-Kjeldhal method for Kjeldhal nitrogen in extraction (Dora S, 1975). Total organic carbon in material of urine was measured by the TOC analyser (SHIMAZU: TOC-5000A).

Results and Discussion

Mass balance

Figure 9.1.2 shows evolution of the total weight of input and sawdust matrix. As seen, total input increased linearly whereas total weight of the matrix was on the stable level around 7 to 8 kg throughout the operation. From the Fig.9.1.2 and Table 9.1.1, it was found that approximately 54kg of input water was successfully removed from the matrix. Figure 9.1.3 shows accumulation of total solid (TS) in the matrix. The accumulation rate of the TS was approximately 29g/day that indicated approximately 60% of input TS decomposed whereas further 40% of input TS remained in the matrix.

Figure 9.1.4 shows the evolution of carbon accumulation in the matrix. While virgin sawdust matrix contained great amount of carbon of 1872g, they hardly decomposed in the composting process (Lopez *et al.*, 2004b). It was therefore assumed that carbon accumulation in Fig.9.1.4 corresponded to the carbon from the toilet waste of daily input: the accumulation rate was approximately 8.76g-C /day. Regarding to the nitrogen, the accumulation rate was 1.82g-N/day (no Figure). The recovery rate of nitrogen was approximately 15% whereas further 85% was lost by ammonia volatilization.

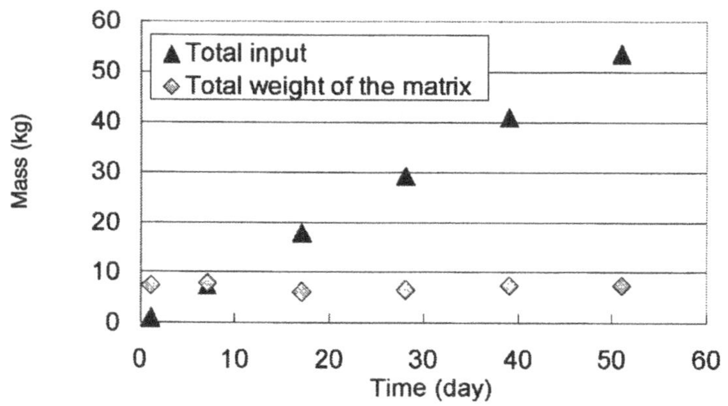

Figure 9.1.2: Evolution of total weight of input and sawdust matrix.

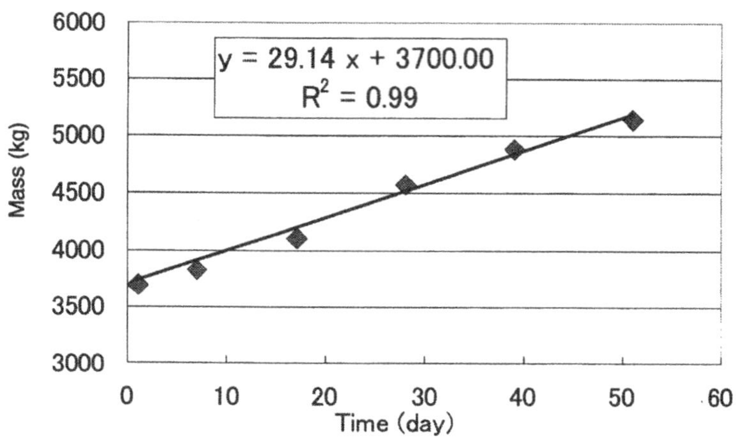

Figure 9.1.3: Evolution of the total solid accumulation.

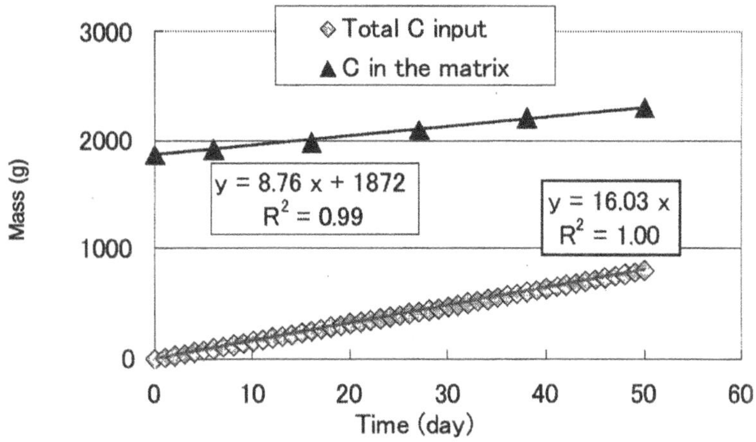

Figure 9.1.4: Evolution of the carbon accumulation.

Power consumption

Table9.1.2 lists the power consumption in the practical operation for 50days. As seen, approximately 80% of total power was consumed only for the heating device. The second biggest consumption was for the ventilation system. In this operation 54kg of input water was evaporated from the matrix.

The latent heat of water at 40°C is 570cal/g and 1kWh is equal to 860kcal. If temperature of whole amount of water in the bio-toilet were at 40°C, required energy for water evaporation would be approximately 31000kcal that could be translated to 35.8kWh. The efficiency for power on the evaporation was therefore estimated to approximately 33% in this operation.

Table 9.1.2: Power consumption for 50days.

Mixing	Heat	Ventilation	Other	Total	Unit
2.2	85	14.4	7	108.6	kWh
2.0	78.3	13.3	6.4	100.0	%

Conclusion

Toilet waste was successfully treated in the bio-toilet for 50days. Sawdust matrix was not replaced to new one during the experimental operation. Approximately 54kg of input water was successfully removed from the matrix by heating device. Constant accumulations of total solid and carbon were found in the bio-toilet: their accumulation rates were 29g-dry/day and 8.76g-C/day respectively. Great amount of the nitrogen (85%) was not recovered by ammonia volatilization. Approximately 80% of power was consumed in the heating device. The efficiency of power consumption on water evaporation was roughly estimated to 33% in this operation.

Acknowledgement

We thank Prof. Minoru Terazawa (Hokkaido University) for his useful suggestions and helps in our experiment. We would like to offer appreciative thanks to Mr. Takashi Kakimoto who donated experimental material. This work has been supported by CREST of JST (Japan Science and Technology Corporation).

Reference

Almeida, M., C., Butler, D. and Friedler, E. 1999. At-source domestic wastewater quality Urban Water System Engineering Water, 1, 49-55.

Dora, S. 1975. Determination of ammonia and kjeldahl nitrogen by idophenol method. Water Research, 10(1), 31-36.

Environmental Technology Association. 1981. Guidebook for domestic wastewater treatment, 20-21.

Hotta, S., Noguchi, T. and Funamizu, N. 2005. Experimental study on nitrogen components during composting process of feces. Environmental Engineering Research, 42, 325-333.

Kitsui, T., Terazawa M., 1999. Environmentally friendly toilets for the 21st century, Bio-toilet. Proceedings of the 10th ISWPC Yokohama, 3, 120-121.

Kempers, A., J. and Zweers, A. 1986. Ammonium determination in soil extracts by the salicylate method. Communications in Soil Science Plant Analysis, 17(7), 715-723.

Lopez, A., M., Funamizu, N. and Takakuwa, T. 2002a. Onsite wastewater differentiable treatment system: modeling approach. Water Science and Technology, 46, 317-324.

Lopez, A., M., Funamizu, N. and Takakuwa, T. 2002b. Characterization of feces for describing the aerobic biodegradation of feces. Journal of the Japan Society of Civil Engineers, JSCE, No.720/VII-25, 99-105.

Lopez, A., M., Funamizu, N., and Takakuwa, T. 2004a. Temperature effect on aerobic biodegradation of feces using sawdust as a matrix. Water Research, 38, 2406-2416.

Lopez, A., M., Funamizu, N. and Takakuwa, T. 2004b. Modeling of aerobic biodegradation of feces using sawdust as a matrix. Water Research, 38, 1327-1339.

Lopez, A., M., Funamizu, N. and Takakuwa, T. 2005. Biological activity in the composting reactor of the bio-toilet system. Bioresource Technology, 96(7), 805-812.

Martins, O. and Dewes, T. 1992. Loss of nitrogenous compounds during composting of animal wastes. Bioresource Technology, 42, 103-111.

9.2 Alleviation of the Effect of Antibiotics on the Aerobic Biodegradation of Feces in Composting Toilet

Takashi Kakimoto[1],*, Naoyuki Funamizu[1]

[1]Graduate School of Engineering, Hokkaido University,
Kita 13 Nishi 8 Kita-ku, Sapporo 060-8628 Japan

(*E-mail: kakimoto@eng.hokudai.ac.jp)
Tel & Fax: 81-11-706-6272

Abstract

We have reported the severe effect of amoxicillin dosage (penicillin group antibiotics) on the feces composting process, and the degradation mechanisms of amoxicillin (the degradation of amoxicillin in a toilet matrix was dependent on the concentration of ammonia, phosphate and hydroxyl ion.) in the composting toilet matrix. Since, in general, to show an inhibitory effect of antibiotics on bacteria, concentration of antibiotics must be more than the concentration that shows the inhibitory effect and its level must be kept for the time needed. This suggested that it is better to keep high phosphate, ammonia, and pH level for rapid degradation of amoxicillin. During the biodegradation of feces, concentration of ammonia and pH level in the sawdust matrix increases due to the decay of organic nitrogen in feces, and thus our objective in this study is to evaluate the alleviation effect of higher feces loading operation. The results indicated that we observed higher enough respiratory activity in higher feces loading operation system than that of ordinary feces loading operation. Therefore we can conclude higher feces loading operation can relieve the inhibitory effect of amoxicillin on the composting process of feces. But the reason why we can observe the relieved bacterial activity seemed not to be the rapid degradation of amoxicillin but to be other reasons like an increase in bacterial counts or its high activity.

Key Words

Amoxicillin, Bacteria, Higher feces loading ratio

Introduction

In the Onsite Wastewater Differentiable Treatment System (OWDTS) (Lopez *et al.*, 2002), household wastewater is separated into three fractions (the blackwater, the higher load graywater, and the lower load graywater) and they can be treated separately. The blackwater is treated by a composting toilet using sawdust as a matrix. The presence of organic matter and such nutrients as N, P, K in human excrement suggests that a stabilized toilet matrix is a good candidate as an effective compost. However, estrogens and a wide variety of pharmaceuticals in human excrement also enter the reactor, which on the other hand have some effects on it.

Among these pharmaceuticals, antibiotics in particular appear to have inhibitory effects on bacteria. One of the most frequently used antibiotics in the world is amoxicillin (Jones *et al.*, 2002; Khan *et al.*, 2004). The effect of amoxicillin in human excrement on the biodegradation process of feces has been reported (Kakimoto *et al.*2006a) and it was found that although the concentration of amoxicillin in toilet matrix declined, we observed severe inhibition on bacteria during the treatment of feces. It has been reported that degradation rate of amoxicillin was dependent on the concentration of ammonia, phosphate and hydroxyl ion concentration in the composting toilet matrix (Kakimoto *et al.* 2006b). Since, among these factors, the concentration of ammonia and hydroxyl ion concentration increased due to the decay of organic nitrogen in feces during the composting process, it may be possible that we can accelerate the degradation of amoxicillin in the matrix by increasing the load on the toilet.

In general, to show an inhibitory effect of antibiotics on bacteria, concentration of antibiotics must be more than the concentration that shows inhibitory effect, and the concentration must be kept for the time needed. Therefore accelerated degradation of amoxicillin would relieve the inhibitory effect on the composting process. Our objective in this study is to evaluate the alleviation effect of higher feces loading operation.

Material and method

Higher feces loading ratio operation

The experiment was conducted with a temperature controlled batch reactor (*ca.*1.9L) in a water bath. Figure 9.2.1 shows the schematic diagram of the bioreactor and the monitoring system. A gas washing bottle was installed behind the bioreactor. The flow rate of the air was controlled by a flow meter at the end of the system. The sawdust used in this study has been prepared for a composting toilet. The ordinary feces loading ratio (F/S ratio) is about 0.5% (g-Feces/g-Sawdust). Thus, the higher feces loading pre-operations in this study were conducted at F/S ratio of 0.5%, 2.5% and 5% for three days. Afterwards, the batch tests (Feces/Sawdust ratio of 0.5%, 2.5% and 5% on dry basis) were conducted after adjustment of the moisture content of the mixture (60%) and the amoxicillin concentration (100 μg/g-dry). While air was supplied from the bottom of the reactor, the oxygen concentration, the temperature, and the pressure of the air were monitored at the inlet and the outlet of the reactor. From these data, we calculated the Oxygen Utilization Rate (OUR) that was used to evaluate the performance of the composting process. According to Lopez Zavala *et al.* (2005), we can say that these experimental conditions conform to the optimum for the feces composting process.

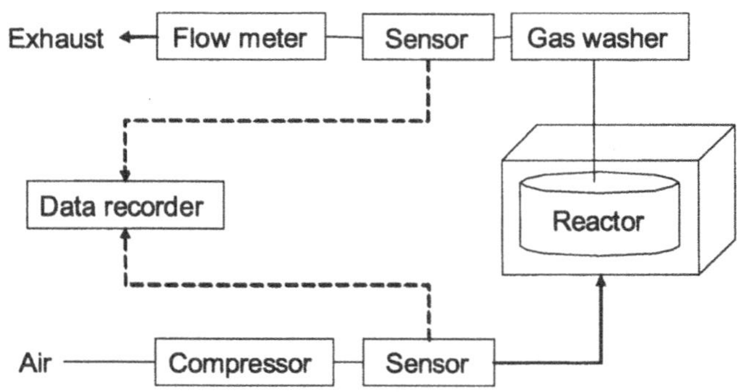

Figure 9.2.1: The experimental device and monitoring system

Determination of amoxicillin in the sawdust matrix

In order to monitor the concentration of residual amoxicillin in the sawdust, we performed extraction and clean up procedures prior to analysis (Kakimoto *et al.*, 2006a). The analyses were performed with a Waters Alliance system equipped with a UV detector (Waters 2487). Sample injections of 10 μl were made with a Waters 717 plus an auto-sampler on an Atlantis® dC18; 3 μm; 4.6 x 150mm column. The separation was performed under an isocratic mode (10 mM phosphate buffer (pH4.8)/Acetonitrile=9/1) at a flow rate of 1ml with detection at 229 nm.

Results and Discussions

Effect of higher F/S ratio operation on the inhibitory effect of composting process

Figure 9.2.2 shows the OUR profiles of three systems with different F/S ratio (0.5%, 2.5% and 5%). As seen in Figure 9.2.2, the 0.5% F/S ratio system seemed to be affected by amoxicillin, and the others showed significantly high respiratory activity. The areas below the OUR profiles were equal to the amount of feces treated during the experiment, and to evaluate the effect of antibiotics on the composting process, we compared the fraction of feces treated in the duration with the total amount of feces input.

Figure 9.2.4 showed the fraction of feces treated in this experiment. The amount of feces treated in this experiment increased depending on the increasing of the F/S ratio. Since the fraction on the operation with 0.5% of F/S ratio was about 30%, it seemed that the 0.5% of F/S ratio was not enough for the stable treatment of feces against the antibiotic dosage. In the case of 2.5% and 5.0% of F/S ratio, 50% and about 90% of feces input was treated, respectively. Thus, we can say that the higher F/S ratio operation alleviate the effect of antibiotics on the composting process of feces.

Concentration of amoxicillin in the sawdust matrix

In general, in order to show an inhibitory effect of antibiotics on bacteria, concentration of antibiotics must be more than the concentration that shows the inhibitory effect, and its level must be kept for the time needed. Thus it seemed that accelerated degradation of amoxicillin would relieve the inhibitory effect. As it has been mentioned above, degradation rate of amoxicillin in composting toilet depended on phosphate, ammonia and hydroxyl ion concentration. Among their concentrations, ammonia and hydroxyl ion concentration will increase by decomposition of organic nitrogen in feces. Therefore, higher feces loading ratio will relieve the antibiotic effect on the composting bacteria.

Figure 9.2.4 showed the change of amoxicillin concentration in the matrix. It was found out that significant reduction of amoxicillin occurred within three hours. As we expected, the higher F/S ratio system showed the higher reduction rate of amoxicillin, but the difference of each reduction rate was not so big. Though we can relieve the effect of antibiotics on the composting process of feces, the relief seemed to be caused not by the accelerated degradation of amoxicillin but by other reasons such as the bacterial counts and their activity in higher feces loading operation system.

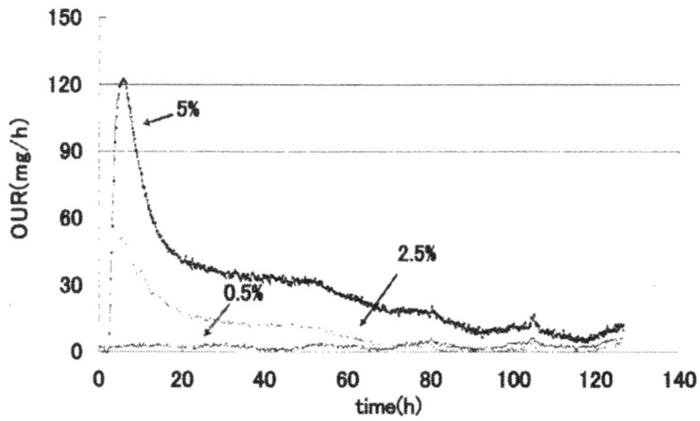

Figure 9.2.2: OUR profiles of three different F/S ratio operations

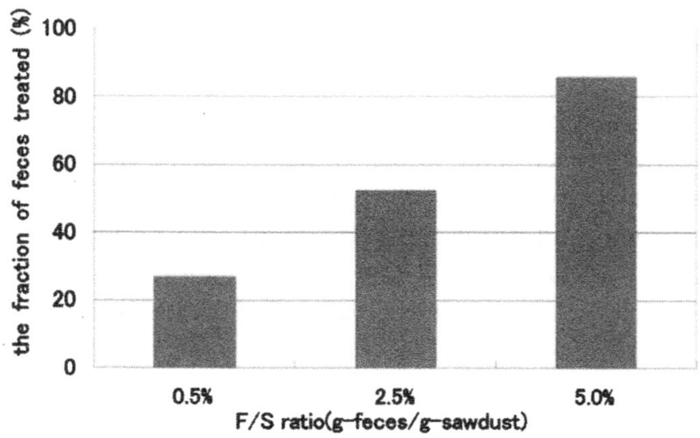

Figure 9.2.3: The fraction of feces treated in this experiment

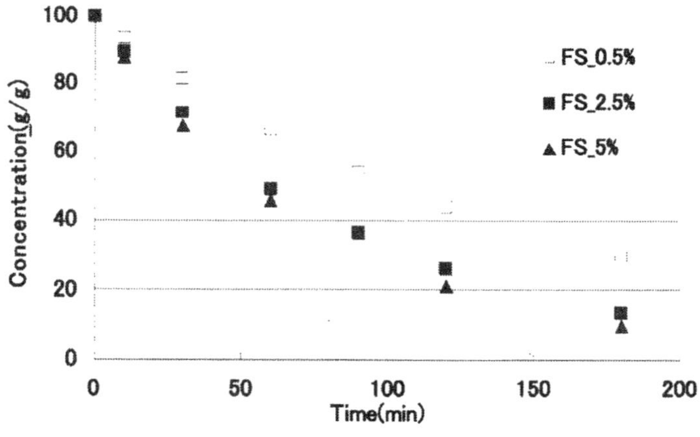

Figure 9.2.4: The concentration of amoxicillin

Conclusions

We examined the effect of higher feces loading operation to relieve the effect of antibiotics on the composting process, and as a result, we obtained the following knowledge.

(1) The higher feces loding operation at F/S ratio of 2.5% or 5.0% can relieve the effect of amoxicillin on the feces composting process.

(2) As we expected, though the degradation rate of amoxicillin increased in higher F/S ratio operations, the rapid reduction of amoxicillin seemed not to be the main reasons for the relief of amoxicillin effect.

Acknowledgement

The authors would like to thank Japan Science & Technology Agency (JST) for the financial support.

References

Jones, O., A., H., Voulvoulis, N., Lester, J., N., 2002. Aquatic environmental assessment of the top 25 English prescription pharmaceuticals. Water Res. 36, 5013-5022.

Kakimoto, T., Osawa, T, Funamizu, N., 2006a. Antibiotic effect of amoxicillin on the feces composting process, and reactivation of bacteria by intermittent feeding of feces. Submitted

Kakimoto, T., Funamizu, N., 2006b. Degradation of amoxicillin in composting toilet using sawdust as a matrix. Submitted

Khan, S., J., Ongerth, J., E., 2004. Modelling of pharmaceutical residues in Australian sewage by quantities of use and fugacity calculations. Chemosphere. 55, 355-367.

Lopez, Zavala, M., A., Funamizu, N., Takakuwa, T., 2002. Onsite Wastewater Differentiable Treatment System: Modeling Approach. Water Sci. Technol. 46, 317-324.

10 Nutrient Management

10.1 Performance Evaluation of a 5-Stage On-Site Nutrient Removal Process

[a]Mohammad J. Mowlaei, [b]Muttucumaru Sivakumar, [c]Dharmappa Hagare

[ab]Sustainable Water and Energy Research Group, School of Civil, Mining & Environmental Engineering, University of Wollongong, NSW, Australia

[c]Principal, Sterling College, Sydney, NSW

[a]Corresponding author: Ph:61 2 4221 5526, Fax:61 2 4221 3238,
E-mail: mj07@uow.edu.au

Abstract

The need to remove nutrients from on-site wastewater treatment systems has received considerable attention in recent years. With increased scientific knowledge derived from laboratory studies and environmental monitoring concerning the impact of nutrients, it is anticipated that a wide variety of on-site treatment technologies will be developed in the near future. This paper examines the performance evaluation of a unique on-site wastewater treatment system that removes the organic constituents as well as nutrients from residential wastewater by biological means. A pilot scale 5-stage treatment system comprising "Anaerobic- Anoxic- Aerobic- Anoxic and Aerobic" stages was investigated in regards to its feasibility for removing Biochemical Oxygen Demand (BOD5), Total Suspended Solids (TSS), Ammonia Nitrogen (NH_3-N), Nitrate Nitrogen (NO_3-N), Total Nitrogen (TN) and Total Phosphorus (TP).The removal efficiency of this 5-stage on-site system was very high with BOD5 removal rate of 95%, TSS removal rate of 97%, NH_3-N removal rate of 88%, TN removal rate of 81% and TP removal rate of 75% in comparison with the conventional aerated wastewater treatment systems (AWTS) of about 50% nitrogen and 40% phosphorus removal. The effluent of this system was also tested to determine its microbiological quality for safe reuse without disinfection. The results indicated that this system is capable of reducing the influent E-Coli drastically with 3.2 log reduction in removal.

Key Words

On-site BNR, Anaerobic, Anoxic, Aerobic, Attached Growth

Introduction

Nutrient control has become a routine part of centralised wastewater treatment plants and for all practical purposes, should become an integral part of on-site wastewater treatment systems. The removal of nutrient and other dissolved and particulate biodegradable constituents in wastewater can be accomplished biologically using a variety of microorganisms, principally bacteria. In attached growth process that has been used in this treatment process the microorganisms responsible for the conversion of nutrients are attached to an inert plastic biomedia namely Accupac-1200. The biological nitrogen removal is a two step process, in the first step which is called nitrification, ammonia (NH_3-N) is oxidised to nitrite (NO_2-N) and subsequently to nitrate (NO_3-N) and in the second step which is called denitrification, nitrate (NO_3-N) is biologically reduced to nitric oxide (NO) and nitrous oxide N_2O and nitrogen gas (N_2) [1]. In the biological removal of phosphorus, the phosphorus in the influent wastewater is incorporated into cell biomass, which subsequently is removed from the process as a result of sludge wasting. The reactor configuration utilised for phosphorus removal is comprised of an anaerobic zone that is placed ahead of an anoxic and an aerobic zone. Under anaerobic conditions, phosphorus accumulating organisms (PAO) will assimilate fermentation products such as volatile fatty acids (VFAs) into storage products within the cells with the simultaneous release of phosphorus from stored polyphosphates. Subsequently under anoxic-aerobic conditions, the energy which is produced by the oxidation of storage products will help to increase the amount of polyphosphate storage within the cells.

Materials and Methods

A pilot scale wastewater treatment system was set up at the Engineering Innovation and Education Centre of the University of Wollongong in Coniston NSW. This pilot scale tank was made from 6 mm thick perspex with the overall dimensions of 2000 mm length, 150 mm width and 300 mm depth which was fed with the residential wastewater collected from Bellambi Sewage Treatment Plant (BSTP) located north of Wollongong. The Biomedia selected for this experiment was Accupac CF-1200 plastic media with specific surface area of 226 m^2/m^3 and void spaces of 95%. This system comprised of 5 zones (anaerobic, anoxic, aerobic, anoxic, aerobic), a pre-sedimentation tank and a final clarifier .
This system was designed to remove nitrogen by aerobic biological nitrification and subsequent denitrification in anoxic zones. The nitrified effluent from the end of the first aerobic zone (3rd Chamber) was pumped back to the head of the first anoxic zone (2nd Chamber), where it was introduced into the incoming wastewater and mixed under anoxic conditions for denitrification. Because the organic substrate in the influent wastewater provides the electron donor for oxidation reduction reactions using nitrate, this process was termed Preanoxic denitrification. The second anoxic zone (4th Chamber) was provided for additional denitrification. This process is generally termed the Postanoxic denitrification as BOD removal has occurred first and is not available to drive nitrate reduction reaction. The Postanoxic denitrification has much slower rate of reaction than for Preanoxic processes using wastewater BOD.

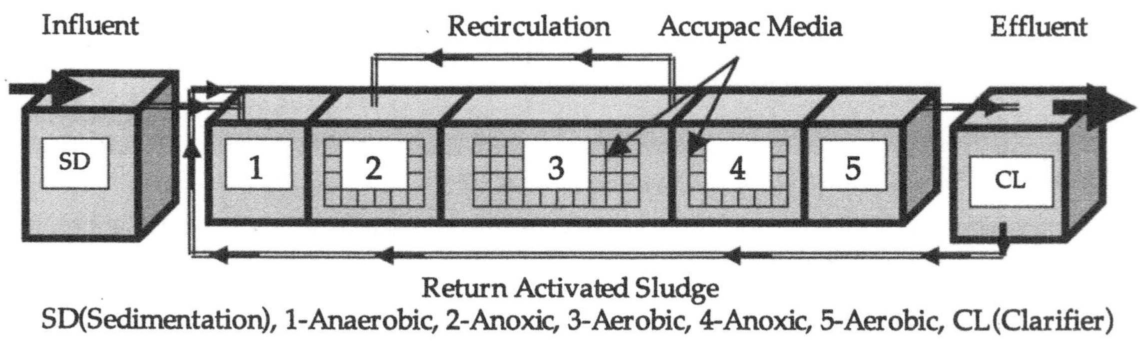

Figure 10.1.1: Schematic of the biological reactor with Pre-Sedimentation and Final Clarifier

In the biological phosphorus removal, the phosphorus in the influent wastewater was incorporated into cell biomass, which subsequently was removed from the process as a result of sludge wasting. Biological phosphorus removal was initiated in the anaerobic zone (1st chamber) and it was removed in the subsequent anoxic and aerobic zones (2nd and 3rd Chambers). Phosphorus accumulating organisms (PAOs) were encouraged to grow and consume phosphorus in systems that uses the reactor configuration of an anaerobic tank ahead of an anoxic-aerobic tank. Under anaerobic condition PAOs will assimilate fermentation products into storage products within the cells with the concomitant release of phosphorus, then under anoxic-aerobic conditions the energy produced will help to increase the phosphorus storage capability within the cells. The last aerobic zone reoxygenated the process flow stream and stripped it of any remaining dissolved nitrogen gas and minimised the release of phosphorus in the final clarifier. A schematic of the system is shown in Figure 10.1.1. This system had the total Hydraulic retention time (HRT) of 27.5 h (approximately one day) which provided sufficient time for sedimentation and degradation of organic compounds. The HRT of each chamber was calculated for better efficiency in nitrification/denitrification process as well as providing an ideal conditions for phosphorus removal. By experiment it was concluded that the HRT of 2h for 1st chamber, 4h for 2nd and 4th chamber, 8 h for 3rd chamber, 1.5h for the 5th chamber and 3.5h for clarifier seemed to provide better effluent quality. The internal recycle of 200 mL/min showed a better result for denitrification process. The return activated sludge of 65 mL/min (equal to flowrate of Q) provided a better source of fermentation product for phosphorus removal compare to the lower flow rate of 1/2 Q. The solid retention time (SRT) of 25 to 30 days seemed to provide enough time for biodegradation of organic compounds and complete nitrification without losing the efficiency.

Results and Discussion

The influent and effluent of the pilot scale WWTS as well as the influent and effluent of each chamber were analysed for various wastewater quality parameters such as Temperature, Dissolved oxygen (DO), pH, Turbidity, Total suspended solids (TSS), Biochemical oxygen demand (BOD_5), Total Kjeldahl nitrogen (TKN), Ammonia-nitrogen (NH_3-N), Organic-nitrogen (Org-N), Nitrate-nitrogen (NO_3-N) and Total phosphorus (TP) over 12 months period from January 2005 to December 2005. Table 10.1.1 shows the variation in influent and effluent wastewater parameters with their minimum, maximum and average values during the sampling period. The results were compared with the data obtained for conventional AWTS [3,4,7] and the primary difference was the higher removal efficiency of nitrogen and phosphorus for the present system. The conventional Aerated Wastewater Treatment Systems (AWTS) use aeration as an integral part of the treatment process and they comprise of 3 principal chambers, "sedimentation, aeration and clarification-disinfection".

Table 10.1.1: Influent and Effluent wastewater quality parameters

Parameters	Present study							Conventional AWTS
	Influent			Effluent			Removal (Average) (%)	Removal (Range) (%)
	Min	Max	Ave	Min	Max	Ave		
BOD (mg/L)	123	307	201	3	24	8	95	80-97
TSS (mg/L)	136	388	236	2	29	5	97	85-98
NH_3-N (mg/L)	17	54	36	2	17	3	88	60- 85
NO_3-N (mg/L)	0	0.8	0.4	2	22	7	-----	-----
TN (mg/L)	26	66	43	4	25	9	81	30- 50
TP (mg/L)	5	17	10	1	8	3	75	25-40
Ecoli. (CFU/100mL)	$5x10^5$	$4x10^7$	$8x10^6$	$7x10^2$	$14x10^3$	$4x10^3$	3.2 log reduction Before disinfection	3-4 log reduction After disinfection

Biochemical Oxygen Demand (BOD_5)

The influent BOD5 varied between 123 mg/L and 307 mg/L, with the average concentration of 201 mg/L while the effluent BOD5 concentration had the minimum of 3 mg/L and maximum of 24 mg/L with the average of 8 mg/L. The average effluent BOD_5 of 8 mg/L was an indication of a significant removal efficiency of this 5 stage treatment system. Figure 10.1.2 shows the BOD_5 removal efficiencies and contribution of each chamber to the removal process. The first chamber (anaerobic chamber) and second and fourth chambers (anoxic chambers) did not contribute to the BOD_5 removal significantly, however as expected the third chamber (aeration chamber) contributed towards substantial BOD_5 removal with the removal rate of 55%. The fifth chamber provided the second highest contribution towards BOD5 removal with the removal efficiency of 50%. The present system showed very high overall BOD5 removal of 95% compared to conventional AWTS of average 88%.

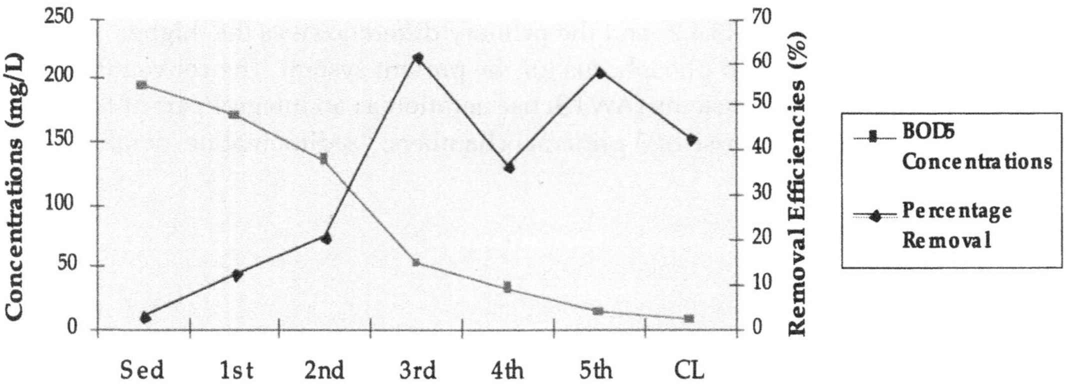

Figure 10.1.2: BOD_5 Concentrations and removal efficiencies in each chamber

Total Suspended Solids (TSS)

The influent BOD5 varied between 123 mg/L and 307 mg/L, with the average concentration of 201 mg/L while the effluent BOD5 concentration had the minimum of 3 mg/L and maximum of 24 mg/L with the average of 8 mg/L. The average effluent BOD_5 of 8 mg/L was an indication of a significant removal efficiency of this 5 stage treatment system. Figure 10.1.2 shows the BOD_5 removal efficiencies and contribution of each chamber to the removal process. The first chamber (anaerobic chamber) and second and fourth chambers (anoxic chambers) did not contribute to the BOD_5 removal significantly, however as expected the third chamber (aeration chamber) contributed towards substantial BOD_5 removal with the removal rate of 55%. The fifth chamber provided the second highest contribution towards BOD5 removal with the removal efficiency of 50%. The present system showed very high overall BOD5 removal of 95% compared to conventional AWTS of average 88%.

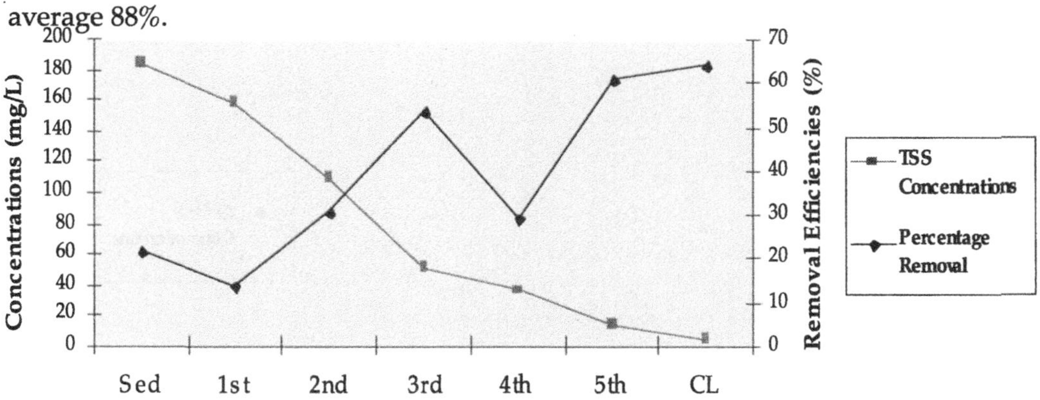

Figure 10.1.3: TSS Concentrations and removal efficiencies in each chamber

Ammonia Nitrogen (NH_3-N)

The concentration of ammonia in the influent varied between 17 mg/L and 54 mg/L with the average concentration of 36 mg/L. The effluent ammonia ranged between 2 mg/L to 17 mg/L with the average concentration of 3 mg/L. The reduction of ammonia from 36mg/L to 3 mg/L was an indication of high removal efficiency of this 5 stage system. Figure 10.1.4 indicate that the highest removal of ammonia had taken place in the 3rd chamber with the removal of 42%. The second highest removal of ammonia occurred in the 5th chamber with the removal efficiency of about 33%. The 2nd chamber had the removal efficiency of 27% while the 4th chamber had the removal rate of 14% only. The clarifier had the removal rate of 13%. The first chamber did not contribute to the removal of ammonia significantly and had the removal efficiencies of about 6%. The sedimentation tank had the lowest ammonia removal with removal efficiency of only 3%. The overall ammonia removal of 88% of the present system was much higher of conventional AWTS of average 70%.

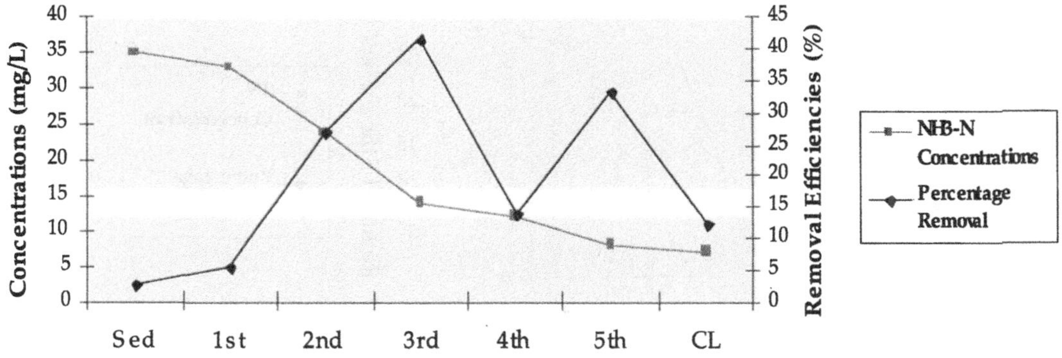

Figure 10.1.4: NH_3-N Concentrations and removal efficiencies in each chamber

Nitrate Nitrogen (NO_3-N)

The comparison of the influent nitrate with the effluent nitrate was an indication of an increase of nitrate in the effluent. Through the biodegradation process with long SRT a large amount of ammonia was oxidated to nitrate in the nitrification process. Hence it was expected that the concentration of nitrate will increase gradually along the various chambers. Figure 10.1.5 indicates that there is been a sharp increase in the nitrate concentration in the 3rd and 5th chamber due to the oxidation of ammonia to nitrate. The highest nitrate production took place in the 3rd chamber with the nitrate concentration of 12 mg/L and the 5th chamber had the nitrate concentration of about 9 mg/L. The removal of nitrate happened mainly in the 2nd and 4th chambers due to denitrification process. The clarifier also contributed to the reduction of nitrate from 9 mg/L to 7 mg/L. The nitrate concentration had a slight increase in the first and second chamber.

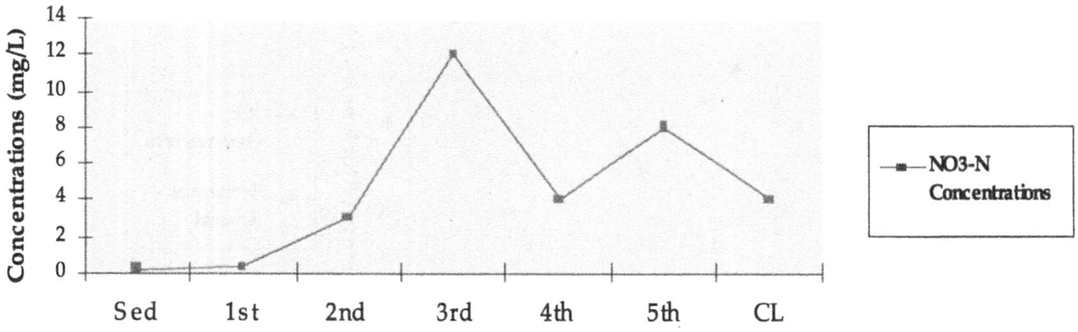

Figure 10.1.5: NH_3-N Concentrations in each chamber

Total Nitrogen (TN)

The influent TN ranged from 26 mg/L to 71 mg/L with an average of 48 mg/L. The average effluent concentration of 9 mg/L was an indication that this treatment system provides conditions for significant nitrogen removal with the overall removal efficiency of 81% compared to conventional AWTS of about 50% nitrogen removal. Figure 10.1.6 demonstrates the removal efficiencies in each chamber. The highest nitrogen removals occurred in the 2nd and 4th chambers (anoxic chambers) due to the effective denitrification process. The clarifier also contributed to the nitrogen removal with the removal rate of 29%. The 3rd chamber had the lowest removal of 7%. The sedimentation tank and the first chamber did not contribute to the nitrogen removal significantly.

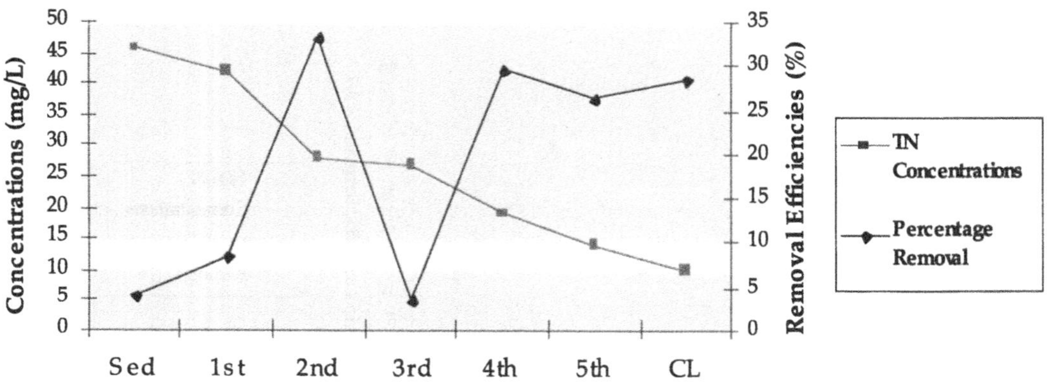

Figure 10.1.6: TN Concentrations and removal efficiencies in each chamber

Total Phosphorus (TP)

The average total phosphorus concentration of 10 mg/L was observed for the influent wastewater which is similar to TP concentration reported for the medium strength untreated household wastewater [1]. The pilot plant produced effluent with TP concentration of 3 mg/L which was higher than the required limit of less than 2 mg/L set by NSW Department of Health [2]. Figure 10.1.7 indicates the existence of more phosphorus in the first chamber due to the release of phosphorus under anaerobic condition however the removal of phosphorus happened in the subsequent chambers. The 2nd and 3rd chambers contributed significantly toward the removal of phosphorus form influent wastewater with the removal rate of 20% and 35% respectively. The 5th chamber also had the removal efficiency of about 20%. The clarifier was responsible for 15% removal. A comparison of the present design removal efficiency of 75% with conventional AWTS of less than 40% revealed that this system can remove a significant amount of phosphorus from the influent wastewater, however modifications are needed to improve the removal efficiency of this pilot tank for phosphorus removal.

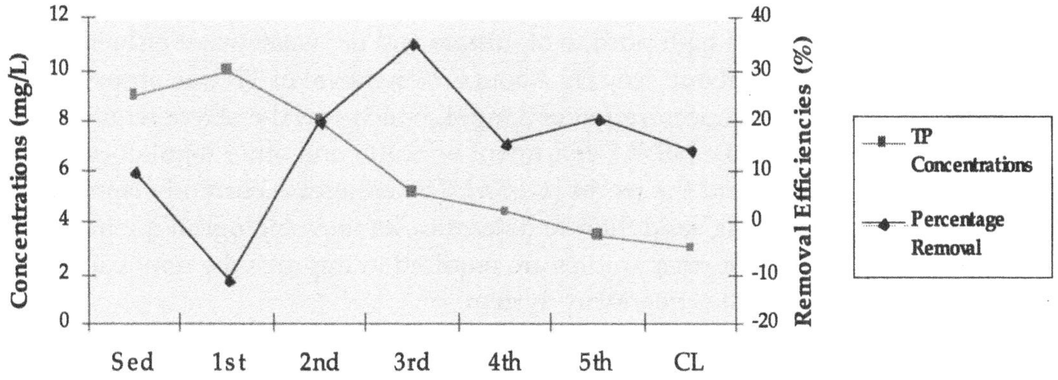

Figure 10.1.7: TP Concentrations and removal efficiencies in each chamber

E-Coli

Escherichia Coliform (E. Coli) is one of the coliform bacteria population and is more representative of fecal contamination. The comparison of effluent and influent E-Coli was an indication of very high removal efficiency of this 5 stage system. This system readily reduced the average influent E.Coli from $8x10^6$ cfu/100mL to the average effluent concentration of $4x10^3$ cfu/100mL without use of any disinfection method. The E-Coli concentration in each chamber had a lot of fluctuation but as a whole the system had 3.2 log reduction in the removal efficiency. Figure 10.1.8 illustrate the concentration and removal efficiency of this system.

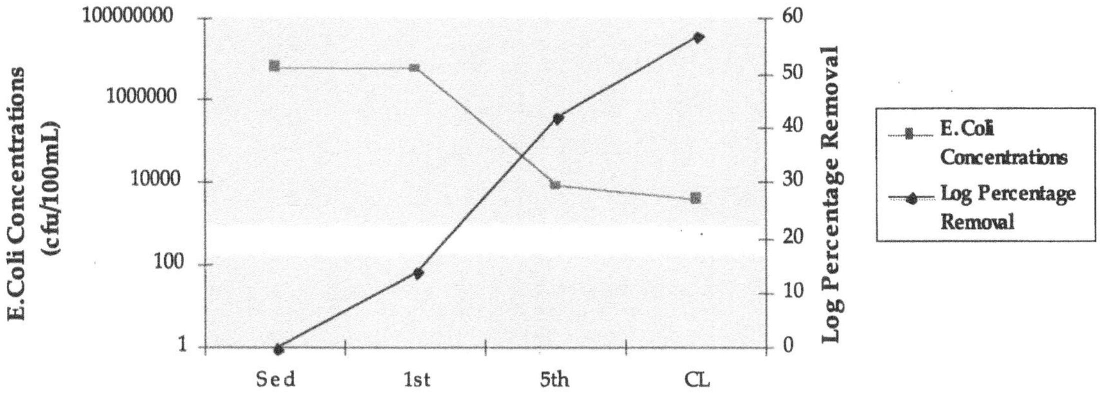

Figure 10.1.8: E.Coli Concentrations and removal efficiencies in each chamber

Conclusion

This study was undertaken to evaluate the performance of a 5 chamber on-site wastewater treatment system for nutrient removal and potential reuse of wastewater for irrigation purposes. A comprehensive investigation in both theoretical and experimental areas revealed that this 5 stage on-site system is capable of removing a high portion of nutrients as well as organic matters. The BOD5 and TSS removal efficiencies of this innovative design were as high as 95%, and 97% respectively with the effluent average concentrations of 8 mg/L of BOD_5 and 5 mg/L of TSS. The results were well below the required limit of 10 mg/L of BOD_5 and 10mg/L TSS for reusable wastewater set by Australian authorities [2,5,6] and American standard reported by Metcalf and Eddy [1]. This on-site system were designed for the primary purpose of nitrogen and phosphorus removal as well as the removal of organic compounds. The nitrogen removal efficiency of 81% was quite satisfactory and the system produced the effluent quality with total nitrogen concentration of 9 mg/L. The reduction of ammonia concentration from average 36 mg/L to 3 mg/L was an indication of effective nitrification process. As expected the nitrate showed an increase in concentration but through the denitrification in anoxic zones it was possible to remove a high portion of nitrate and the wastewater effluent had the nitrate concentration of about 7 mg/L. About 75% removal of TP was observed yielding an average effluent TP concentration of 3 mg/L which was the above required level of less than 2 mg/L set by the NSW Department of health and other regulatory authorities in Australia and around the world [1,2,5,6]. The effluent is currently being tested passing through a polishing sand filter to determine its microbiological quality for safe reuse without disinfection. Further studies are required to improve the removal efficiency of this on-site wastewater treatment system.

Acknowledgements

The authors wish to thank the Sustainable Water and Energy Research Group (SWERG) for supporting this project and Mr. Norm Gal and Ms Joanne Gorge, the Environmental Engineering Laboratory Technical staff for providing technical assistance to this project.

References

Metcalf & Eddy, Inc. (2003) Wastewater Engineering, Treatment and Reuse, 4th edition, McGraw Hill, New York, USA

NSW Department of Health (2001) Domestic Wastewater Treatment Systems

Accreditation Guidelines. NSW Department of Health, Australia

EPA Victoria (2002) Guidelines for Aerated On- Site Wastewater Treatment Systems.

Environmental Protection Authority, Victoria, Australia. Publication 760.

U.S. EPA. (2001) On-site Wastewater Treatment and Disposal Systems. Technical

Report EPA 625/1-01-102. Washington, D.C., U.S. Environmental Protection Agency.

NSW Department of Local Government (1998) Environment and Health protection Guidelines, On-site sewage management for single households, Bankstown, NSW, Australia

AS/NZS 1547.2000 On-site domestic wastewater management, Part 4, Wastewater treatment units, Standard Australia/ Standard New Zealand

Martin, B. (2001) The operation and performance of the Bioseptic AWTS. On-site 01 conference. University of New England, pp271-278.

11 Water and Wastewater Management

11.1 A Technology Database for Decentralised Wastewater Recycling Systems

Shaun Jamieson, Martin Anda and Goen Ho
Environmental Technology Centre, Murdoch University Western Australia.
Email: A.Anda@murdoch.edu.au

Abstract

There are many wastewater treatment and reuse technologies available for small-scale, decentralised and integrated applications. Selection of the most appropriate technologies should be made to ensure public health and environmental quality is protected, while also considering possible construction limitations and ongoing management and maintenance issues. Choosing from the large range of onsite wastewater treatment systems can therefore be complex and time consuming, which may cause the involved parties to disregard a decentralised approach, or alternatively select inappropriate technologies. In order to make the technology selection process easier a Decentralised Wastewater Treatment Technology Database has been developed. It includes a range of commercially available products from both Australia and overseas in the following categories:

- Aerobic treatment units;
- Sand, soil and peat filters;
- Composting systems;
- Ponds and constructed wetlands;
- Anaerobic digestion;
- Physico-chemical systems;
- Advanced filtration;
- Urine separation and treatment; and others.

The various products in each category are grouped into subcategories based on the different treatment processes. The categories and subcategories are provided with process description, comparative evaluation and suggestions for application. The database helps to provide a more straightforward and simplistic means of decentralised wastewater system technology choice, and can be used by land owners, developers and associated governing bodies to reduce the complexity associated with developing decentralised wastewater systems.

Key Words

Database; decentralised wastewater; technology selection

Introduction

With the continuing increase in water demand and declining rainfall (Berti *et al.*, 2004) it is clear that many Australian cities and towns must move towards more sustainable water management strategies. A major way in which this can be achieved is by implementing decentralised systems that collect and treat the wastewater on a much smaller scale than the traditional approach. The localised nature of the systems means that the treated effluent can be reused for applications such as irrigation, toilet flushing or washing water.

While decentralised systems present a range of benefits, selecting the most suitable approach can present a major hurdle to the involved designers due to the vast array of system configurations available (Hellstrom and Jonsson, 2003). To simplify the overall technology choice process an enhanced wastewater technology database was developed that incorporates a range of conventional and alternative commercially available technologies, an extensive cataloguing system to group the product, description and comparisons on the technology types as well as suggestions for application.

The need for a new database

Selection of the most appropriate technologies should be made to ensure public health and environmental quality is protected, while also considering possible construction limitations and ongoing management and maintenance issues (IETC, 2001). To realize this, a general system configuration must initially be determined, which poses the questions:

- Do we implement source separation (e.g. greywater or urine separation)?
- What scale of collection should be employed (i.e. individual house or multiple connections)?
- What technology types should be used? and
- What will be the end use of the treated wastewater?

However, scientific and case study information on the various technical components involved can be hard to obtain, often resulting in a complex and time consuming decision making process.

Once the general wastewater recycling system characteristics have been agreed upon the designer will usually indiscriminately select a number of wastewater treatment product suppliers to detail their proposed solution. This haphazard process can often mean that not all possible options are considered and the end decision is reliant upon information provided by the supplier. This can potentially cause a disregard for wastewater recycling due to uncertainty, or produce an end decision that is largely based on capital cost rather than a full analysis of the technology being proposed.

A comprehensive wastewater technology database could significantly reduce the difficulties faced by potential wastewater recycling system implementers, by providing a library of wastewater solutions (Green and Ho, 2005). It should not only provide a list of categorised products, but also help to educate the user to the various processes, characteristics and potential applications of the identified technology types. This would help the involved parties to make a more informed and science based decision.

While some wastewater technology databases already exist they were found to be inadequate in a number of areas, including:

- They usually include a limited range of technology types (i.e. Aerobic Treatment Units only). This does not allow the database user to investigate and consider all the possible options available.

- They have an inadequate indexing system to group the different products. This can cause the product database to be confusing and overwhelming.
- They provide minimal information on the identified technologies to which a constructive comparison can be made. Without technology or product comparison the database will have a limited use.

Decentralised Wastewater Treatment Technology Database

A new Decentralised Wastewater Treatment Technology Database was developed at the Murdoch University Environmental Technology Centre, with an aspiration to provide a tool that will assist the implementation of wastewater recycling. In order to address the issues discusse earlier the new database was developed with a number of advanced characteristics. These characteristics have been included to provide a wastewater recycling technology database that will educate the user to the various options available, and then provide a means of comparing those systems. These characteristic include:

1. A range of conventional and alternative commercially available technologies
2. An extensive cataloguing system to group the products.
3. Description and comparison on the technology types
4. Suggestion for application

1. Conventional and alternative commercially available technologies

The database aims to include all possible wastewater treatment and management systems and processes available. This includes the conventional treatment processes such as aerobic treatment units, constructed wetlands and greywater systems, as well as a range of alternative approaches such as low-flush toilet systems, composting toilets, urine separation and anaerobic biogas production.

The database also embraces technologies that are applicable on all types of domestic wastewater situations, from the single home to large residential developments. This is referred to in the database as 'scale of collection' and involves three potential options of lot (single home), cluster (small to medium groups of homes) or village (large residential developments). The general parameters for each scale are detailed in Table 11.1.1 (below).

Table 11.1.1: The three scales of collection included in the database

Indicated wastewater collection size	Wastewater treatment scale
1 home	Lot
Less than 50kL/day	Cluster scale
More than 50kL/day	Village scale

2. Cataloguing system

The various technology types identified for the database were grouped into separate categories so that the user can easily distinguish between the different products. The categories were firstly split into two major groups including core treatment types and additional components (Table 11.1.2, below). The core treatment types are technologies that are designed to provide the main (usually secondary treatment level 20mg/L BOD, 30mg/L SS). The additional components include any technologies that will provide supplementary treatment or are required in addition to certain core treatment processes. The core treatment type section is divided into seven major categories and the additional components section is divided into six categories (Table 11.1.2, below).

Table 11.1.2: The technology type categories of the new database

	Technology type categories
Core treatment types	1.0 Aerobic treatment units
	2.0 Soil, sand and peat filters
	3.0 Composting systems
	4.0 Ponds and wetlands
	5.0 Anaerobic treatment
	6.0 Physio-chemical treatment
	7.0 Greywater treatment
Additional components	A-1 Toilet systems
	A-2 Primary treatment systems
	A-3 Urine treatment
	A-4 Advanced filtration
	A-5 Disinfection units
	A-6 Additives and household products

Table 11.1.3: The technology type subcategories identified for each category of the core treatment types.

Group	Category	Subcategory
1.0 Aerobic Treatment Units	1.1 Suspended growth	1.1a Activated sludge (continuous aeration)
		1.1b Activated sludge (intermittent aeration)
		1.1c Membrane bioreactor
	1.2 Attached growth (forced aeration)	1.2a Submerged aerated filter (continuous aeration)
		1.2b Submerged aerated filter (intermittent aeration)
		1.2c Fluidised bed
		1.2d Moving bed bioreactor
	1.3 Attached growth (passive aeration)	1.3a Percolating Filter (Primary = settling)
		1.3b Percolating Filter (Primary = humus)
		1.3c Rotating biological contactor
2.0 Soil, sand and peat filters	2.1 Amended soil filter	
	2.2 Peat bed filter	
	2.3 Sand filter	
3.0 Composting systems	3.1 Wet composting	3.1a High influent (liquid separation)
		3.1b Low influent (heater)
	3.2 Dry composting	3.2a Remote (no heater)
		3.2b Remote (heater)
		3.2c Self contained (heater)
4.0 Ponds and wetlands	4.1 Constructed wetlands	4.1a Free water surface
		4.1b Subsurface water flow
	4.2 Ponds	4.2a Aerobic
		4.2b Facultative
5.0 Anaerobic systems	5.1 Open digesters	
	5.2 Closed digesters	
6.0 Physico-chemical	6.1 Electro chemical	
	6.2 Dissolved air flotation	
	6.3 Screening and membrane	
7.0 Greywater treatment systems	7.1 Diversion systems	
	7.2 Secondary units	
	7.3 Constructed wetlands	
	7.4 Advanced systems	

The categories (Table 11.1.2) were dived into a number of subcategories to provide further refinement of the technology types (Table 11.1.3). This cataloguing system will enable the user to identify the general treatment processes for each product simply by the categories and subcategories they are listed under. So far 190 products from both Australia and beyond have been entered into the database within their appropriate categories, with the potential to continually expand this number.

3. Description and comparison

Description and comparison are provided in the database that can be analysed by the user to help in the technology selection process. Rather than provide this information about each individual product the database focuses on the technology types identified in the cataloguing system. Once the user has determined which technology type or types are the most appropriate they can independently analyse the suitable products by contacting the suppliers.

General descriptions are provided for each technology type to explain the processes to the user. The descriptions are designed to start progressively with elementary information (e.g. how an aerobic treatment system works in general) to more specific detail.

An evaluation system is used to compare the technology types within the database. The system comprises of a numbering matrix that provides a score between one and ten (one being the best score) for a range of characteristics called 'evaluation criteria'. There are seven evaluation criteria which include;

i. **Organic (performance):** likely BOD and SS in treated effluent

ii. **Nutrient (performance):** likely nutrient (NH_4, TN and TP) removal

iii. **Capital cost:** approximate capital cost indication

iv. **Energy use:** likely energy use in kWh/kL

v. **Management cost:** approx. maintenance or service contract costs per year

vi. **Footprint:** total land area required

vii. **Sludge:** Volume of sludge required for treatment/disposal

The evaluation scores are provided for each of the core treatment type categories (including subcategories) and for each available scale of collection for that category (i.e. lot, cluster or village). The scores are based on real values, which were formulated using a combination of sourced literature and the authors' assessment. Four of the evaluation criteria vary significantly between the scales of application indicated. For cluster or village applications these values are based on the 'rated daily volume' (RdV) of the treatment plant (i.e. how many kilolitres of wastewater the system is designed to receive and treat daily - Table 11.1.4, below), while lot scale systems are based on per household values.

Table 11.1.4: An example of the indicative values and correlating scores for each evaluation criteria.

Evaluation criteria	Description	Value	Score (example)
Organic (performance)	BODmg/L / SSmg/L	20/20	5
Nutrient (performance)	Ammonia (% removal)	>80	5
	Nitrogen (% removal)	<60	
	Phosphorus (% removal)	<35	
Energy use	kWh/kL	1.5	8
Capital cost	$/RdV	5,000	7
Management cost	$/RdV/year	200-300	6
Footprint	m^2/RdV	0.6-1.25	4
Liquid Sludge	(kL/RdV/year)	10	6

** RdV – Rated daily volume (kL/day)*

The comparison system is intended to provide basic level of technology type contrasting and may not always reflect the exact characteristics of every product within that category, which is explained within the electronic tool. A prime example being the organic and nutrient impact scores (and values), which can be improved using system modifications or adding additional components, such as biological nutrient removal or membrane filtration. These extra components are likely to increase other evaluation criteria (i.e. capital cost, energy use, footprint, etc.); however, currently the database does not include this assessment.

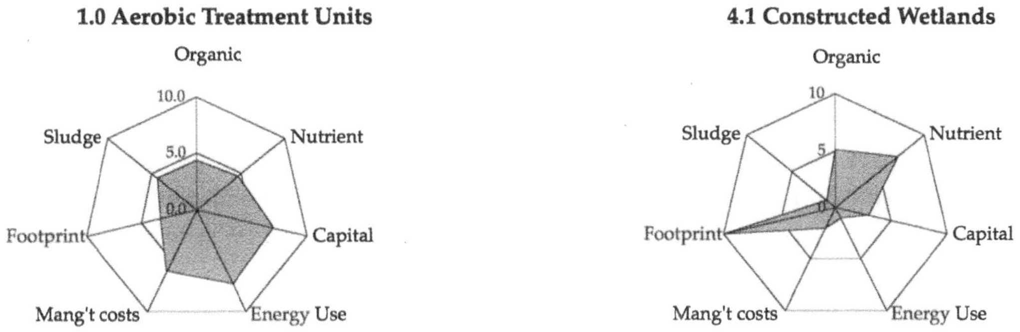

Figure 11.1.1: An example of the scores being used in radar graphs to provide a visual comparison.

The evaluation system results are displayed in the database as radar graphs in order to provide a visual comparison for the evaluation criteria (Figure 11.1.1). The example provided in Figure 11.1.1 includes a general comparison between aerobic treatment units and constructed wetlands. The graph very quickly demonstrates the lower capital cost, energy use and management costs of constructed wetlands, but at the expense of a much higher footprint.

4. Suggestions

Various technology types are provided with suggestions for application. For example: "Membrane filtration is used to achieve a high quality effluent that can typically be used for irrigation with uncontrolled public access or third pipe systems". Each product within the database is also provided with recommended scales of application (i.e. lot, cluster or village).

Benefits and further development

The new Decentralised Wastewater Treatment Technology Database provides a comprehensive repository whereby individuals can source and compare a range of wastewater treatment technologies and products. This helps to remove the complexities associated with choosing the right wastewater treatment approach by:

- Allowing for a quick summarisation the available technologies, to which the user can then begin a simple process of elimination to narrow the options down based on local conditions and user preference.
- All avenues of possible wastewater treatment can be considered (e.g. from constructed wetlands to urine separation)
- It provides a description of the different technology type categories, which will help to educate the users on wastewater treatment processes

- By providing the user with a better understanding of the systems, they can be more confident in making a technology selection or recommendation
- The manufacturers contact details will be displayed, making further information gathering quicker and easier
- The database can be easily updated to include new technologies and remove systems that are no longer available
- The level of information provided is likely to be useful for a range of parties, from novice single home builders to professional consultants and engineers.

The database still requires some editing and enhancement before it can be made publicly available. An electronic decision support tool (EDST) is currently being developed at the Environmental Technology Centre, which will accompany the database. The EDST will aim to further reduce the complexity of appropriate wastewater technology choice by asking a series of questions to determine the social and environmental characteristics, as well as the user preferences to determine the suitable technology types to be employed.

We are trialling the application of the data base and EDST in a number of land developments.

Conclusion

Selection of the most suitable small scale wastewater treatment system can be time consuming and complex, partly due to wide variety of systems and processes available. This can ultimately deter the employment of this more sustainable approach, or alternatively cause an incorrect technology choice, potentially resulting in bad social, economic and environmental experiences.

In order to make the incorporation of decentralised wastewater treatment systems more accessible a unique and innovative technology database was developed at the Murdoch University Environmental Technology Centre, which can provide a range of benefits to planners, consultants and other involved parties. As well as providing a comprehensive cataloguing of a large variety of wastewater management technologies, the database also provides a means of user education and technology comparison. It is likely that the database will not only aid the selection of the most appropriate wastewater technology, but also help the involved personnel to understand and be confident in the end decision.

References

Berti, M.L., Bari, M.A., Charles, S.P. and Hauck, E.J. (2004). *Climate Change, Catchment Runoff and Risks to Water Supply in the South-West of Western Australia*, Department of Environment, Government of Western Australia.

Green, W. and Ho, G. (2005). Small scale sanitation technologies. *Water Science Technology: Onsite Wastewater Treatment, Recycling and Small Water and Wastewater Systems*. **51** (10): 29-38

Hellstrom, D. and Jonsson, L. 2003. Evaluation of small wastewater treatment systems. *Water Science Technology; Small Water and Wastewater Treatment Systems V*. 48 (11-12): 61-68.

International Environmental Technology Centre (IETC). 2001. *Environmentally sound technologies in wastewater treatment for the implementation of the UNEP global programme of action (GPA) – "Guidance on Municipal Wastewater"*. IETC Technical Publication Series [21] – Version 1.

11.2 The Impact of Limiting Factors in a Recirculating Wastewater Treatment System

Ben Kele*, David J Midmore*, Keith Harrower*, Barry Hood*

* Plant Sciences Group, Central Queensland University,
Rockhampton Qld 4702 Australia

E-mail: b.kele@cqu.edu.au, d.midmore@cqu.edu.au,
k.harrower@cqu.edu.au, b.hood@cqu.edu.au

Abstract

The Central Queensland University (CQU) has developed a recirculating evapotranspiration channel for the treatment and reuse of wastewater. The major advantages of the system are an independence from the local soil type and the separation of effluent from external environment. The channel is filled with a soil blend that is suitable for the reuse of effluent, and the sub-surface irrigation in a sealed channel isolates the wastewater and protects public and environmental health. Within the channel the normal limiting factors in on-site wastewater treatment and reuse, such as salinity, sodicity, heavy metals, and nutrient accumulation do have an impact. The extent of impact within the channel was examined over eight trial sites in Central Queensland. The results show that the characteristics of the wastewater had a major influence on how the limiting factors affected the soil within the channel. The examination of the limiting factors within an enclosed environment has produced useful data for the determination of the most beneficial soil blends for on-site wastewater treatment and reuse.

Key Words

Cation exchange capacity, chemical toilet dump, salinity, sodicity, standards, sustainability

Introduction

The Australian and New Zealand Standard 1547:2000, and most related state legislation asks that on-site treatment systems have a sustainable life of at least 15 years (AS/NZS 2000). However no definition of what is an unsustainable system and the associated contaminants of concern exists within the standard or associated legislation. This leaves open the question of when does a system become unsustainable and how is this monitored? The AS/NZS 1547:2000 is a code designed to apply to systems of ten equivalent people (EP) or less; although this has been increased to twenty EP in the state of Queensland. All systems above 20 EP in Queensland need to be licenced by the Environment Protection Agency (EPA) and be regularly tested for characteristics such as faecal coliforms, biochemical oxygen demand (BOD5), total nitrogen (TN), total phosphorus (TP), dissolved oxygen (DO), total suspended solids (TSS), and pH. While these are standard wastewater analyses the link between them and a sustainable on-site wastewater treatment technology is not firmly established, especially when considering the disposal area for the treated effluent or potential recycling uses (Patterson 1997). Many standard wastewater tests were developed to address the environmental and public health concerns relating to the disposal of treated effluent to waterways, and therefore may not identify the treatment sustainability of on-site wastewater technologies where waterway disposal is not used (Han, Kingery *et al.* 2000). However, what are the major environmental or public health concerns within a treatment system or its disposal or reuse area? The majority of on-site systems in Australia apply treated effluent to a designated land/soil area (Beal, Gardner *et al.* 2005; Gardner 2005). The adverse impacts of heavy metals, salinity, sodicity, sodium absorption ratio (SAR; for water), increased sodium percentage of cations (ESP; for soil) and the coupled decrease in cation exchange capacity (CEC; for soil) have all been identified as potential limiting factors in the sustainable application of treated effluent to soil (Siebe 1995; Bond 1998; Graaff and Patterson 2001).

The Central Queensland University (CQU) has developed a recirculating evapotranspiration soil-filter bed on-site treatment technology called the Kele effluent wastewater treatment (KEWT) system (Kele 2005). The recirculatory nature of the KEWT system allows for the possibility of potential limiting factors to accumulate within the recycled effluent and within the soil placed into the channels. Therefore long-term analysis of both the effluent and the soil within the KEWT systems was undertaken for both standard wastewater characteristics and potential limiting factors. This paper describes a selection of the potential limiting factors results from three trial sites, a caravan park (Gem-Air site), a concrete factory (Rockhampton site), and a toilet amenities block incorporating a chemical toilet dump (Sapphire site).

Methods

Salinity accumulation within the recycled effluent

A long-term salinity accumulation study was conducted, in which the holding tank effluent, containing a mixture of new aliquots of primary treated wastewater inputs and recycled water, was examined. Two TPS™ meters were used to establish pH, salinity, DO and temperature. The first meter a TPS™ WP81 measured pH, EC (μS), TDS (ppm), while the second meter a TPS™ WP90 measured DO (ppm) and temperature (°C). Prior to leaving CQU the TPS meters™ were calibrated according to the methods described by the manufacturer. The holding tank at each site was examined once every three-months. A Van-Dorn sample collection device was used to take samples at the top, in the middle, and 500 mm off the bottom of the holding tanks. A Van-Dorn sample collection device can be triggered to take samples at selected depths. Samples were taken at different depths to control for any stratification of effluent that may have occurred within the holding tanks. Three samples were taken at each depth and placed into a 200 ml container that had a depth of at least 100 mm. The two TPS™ meters were used to test the samples at the site. Probes were inserted to a minimum effluent depth of 80 mm and were left within the effluent until a stable reading was achieved. Probes were washed with distilled water between each test. Samples were tested within 20 minutes of being collected. The results from each sample were recorded and, are presented as the average readings of the total number of tests performed in each holding tank for each parameter.

Salinity leaching trial

A leaching trial examined the impact on salinity that tank pump-outs and a 3000 L freshwater flush had on the Rockhampton KEWT system. The trial occurred on the first Wednesday in November 2003. This was after the long-term accumulation trial had been completed. Eight hours worth of tests were performed on the effluent from eleven samples points.

These sample points were:

1. Untreated greywater – taken from the top of the vertical greasetrap
2. Treated greywater – taken from the discharge point of the vertical greasetrap
3. Septic – taken from the discharge point of the septic tank
4. Holding tank – taken from 500 mm off the bottom of the holding tank
5. Non-transpired effluent – taken from the return line from the KEWT channels
6. Bamboo WS – KEWT channel planted with bamboo closest to workshop
7. Bamboo RD – KEWT channel planted with bamboo closest to the road
8. Heliconia – KEWT channel planted with Heliconias (ornamental gingers)
9. Citrus – KEWT channel planted with citrus
10. Banana - KEWT channel planted with banana
11. Pump-well – taken from 500 mm from the bottom of collection tank for the non-transpired effluent prior to it being pumped back to the holding tank

In between the 4th and the 5th hours the septic tank, vertical greasetrap, holding tank and pump-well were all pumped out so that only a 1/4 of the previous volume remained. This volume of effluent was kept to help ensure that the microorganism cultures that had developed in the various treatment stages of the KEWT system were retained. The septic tank and the vertical greasetrap were then filled to just below their respective effluent discharge points with reticulated town water. Then 3000 L of reticulated town water was added to the holding tank. While this water was added the pump was engaged and water was continually pumped through the KEWT channels. After approximately 2000 L of reticulated water had been added to the holding tank a mixture of reticulated water and non-transpired effluent was returned to the holding tank through the return line from the pump-well. The pump-out of the tanks and the addition of the 3000 L of reticulated town-water took approximately 40 minutes. The pump in the holding tank was kept running until the end of the hour. Samples were then taken once an hour for the next fours hours from all 11-sample points and analysed using the methods and equipment previously described. The main aim of this experiment was to test whether accumulated salts could be leached out of a KEWT channel system.

Accumulation of selected limiting factors within the KEWT channel soil

The soil imported into the channels at each site was characterised prior to effluent application. A composite soil sample was obtained from four sections of the channel at the time of installation. Soil core samples were obtained from all sites and transported and stored at 4°C. The Incitec laboratories (NATA accredited) on Gibson Island in Brisbane performed all analysis. Sodium was determined through an ammonium acetate reaction and measured on an ICP AES (Incitec 2003). Chloride was prepared and measured colormetrically in a segmented flow analyser (Incitec 2003). Electrical conductivity (EC) was measured with a conductivity meter. The EC of saturated extract (se) is based on conversions of EC (1:5) and soil texture class, to obtain a more meaningful determination of the soil salinity hazard (Incitec 2003).

The equation used to calculate the cation exchange capacity was:
Cation Exchange Capacity = the sum of K, Ca, Mg, Na, and Al (Incitec 2003)

The equation used to calculate the sodium% of cations was:
Exchangeable Sodium Percentage = Na x 100/CEC (Incitec 2003)

Results

Figure 11.2.1 shows the data on salinity for the holding tank effluent at the Sapphire site. This site had wastewater inputs from a toilet block and a chemical toilet dump point.

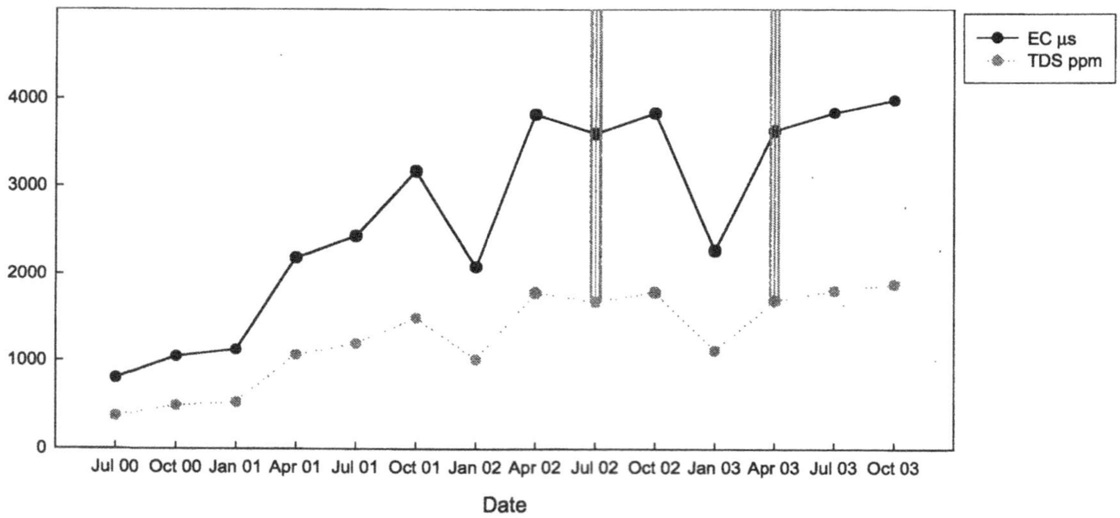

Figure 11.2.1: Salinity long-term accumulation study in the holding tank at the Sapphire site.

In January 2002 and January 2003 wet weather events filled the emergency soakage drain and stormwater infiltrated the holding tank and lowered the EC concentrations of the effluent. The volume of stormwater that entered the holding tank could not be calculated. Within six-months of the site installation the EC concentrations had accumulated to between 3000 and 4000 ÌS. Salts accumulated in this holding tank effluent at unsustainable concentrations.

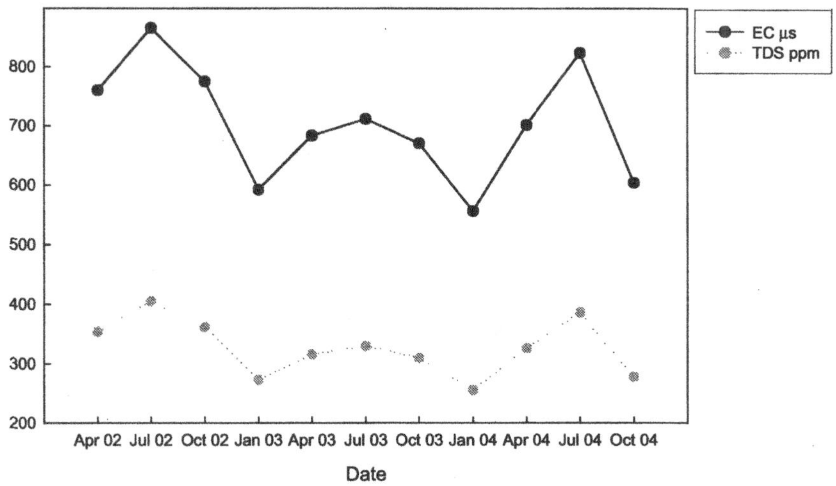

Figure 11.2.2: Salinity long-term accumulation study on the effluent in the holding tank at the Gem-Air caravan park.

The KEWT channel system at Gem-Air caravan park, this was not designed as a closed recirculatory mechanism (such as the Rockhampton or Sapphire sites); but rather diverted treated effluent into a separate tank for reuse in the gem mining industry. The salinity concentrations peaked in the tourist season and decreased dramatically in the warmer non-tourist months. The salinity concentrations (Figure 11.2.2) did not increase in the 2003 tourist season to the same extent as in the other two years. In 2003 the entire area was drought declared and water was transported from Emerald (≈ 80 km). This water had a lower quantity of salts (≈ 350 μS) then the groundwater normally used at Gem-Air (≈ 500). The salinity did not increase to the same extent in 2003 due to the smaller quantities of salt introduced into the wastewater.

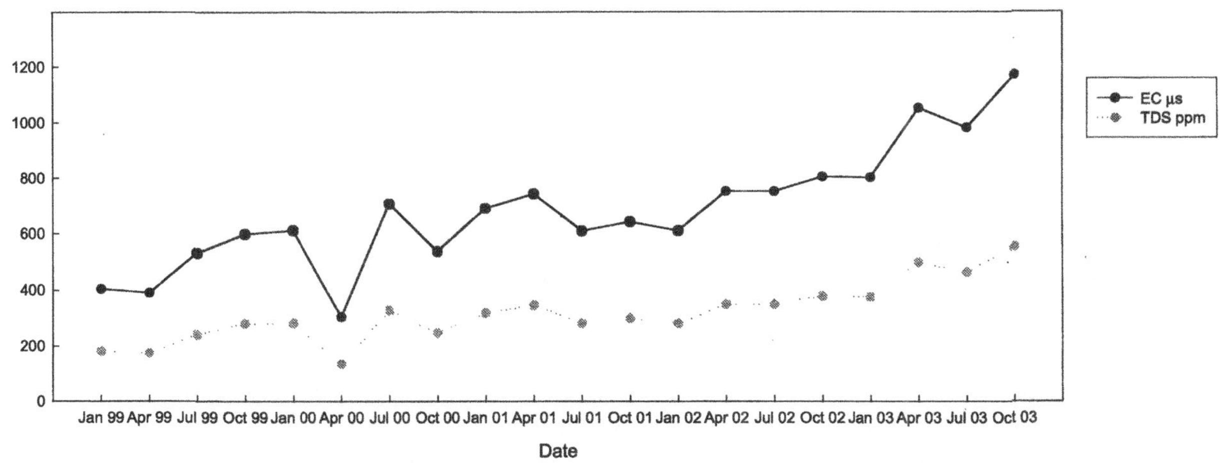

Figure 11.2.3: Long-term salinity on the effluent in the holding tank at the Rockhampton site.

Salinity concentration accumulated over time in Rockhampton (Figure 11.2.3), although there was a reduction in April 2000. At this time a float valve in a toilet at the site malfunctioned which added approximately 6000 L of reticulated town-water to the system. Salinity concentration within the system decreased because some of the excess water that was added to the system was disposed of through the emergency overflow, allowing salts to exit and not be recirculated.

Data on the effluent EC in the KEWT system during the leaching trial are presented in Figure 11.2.4.

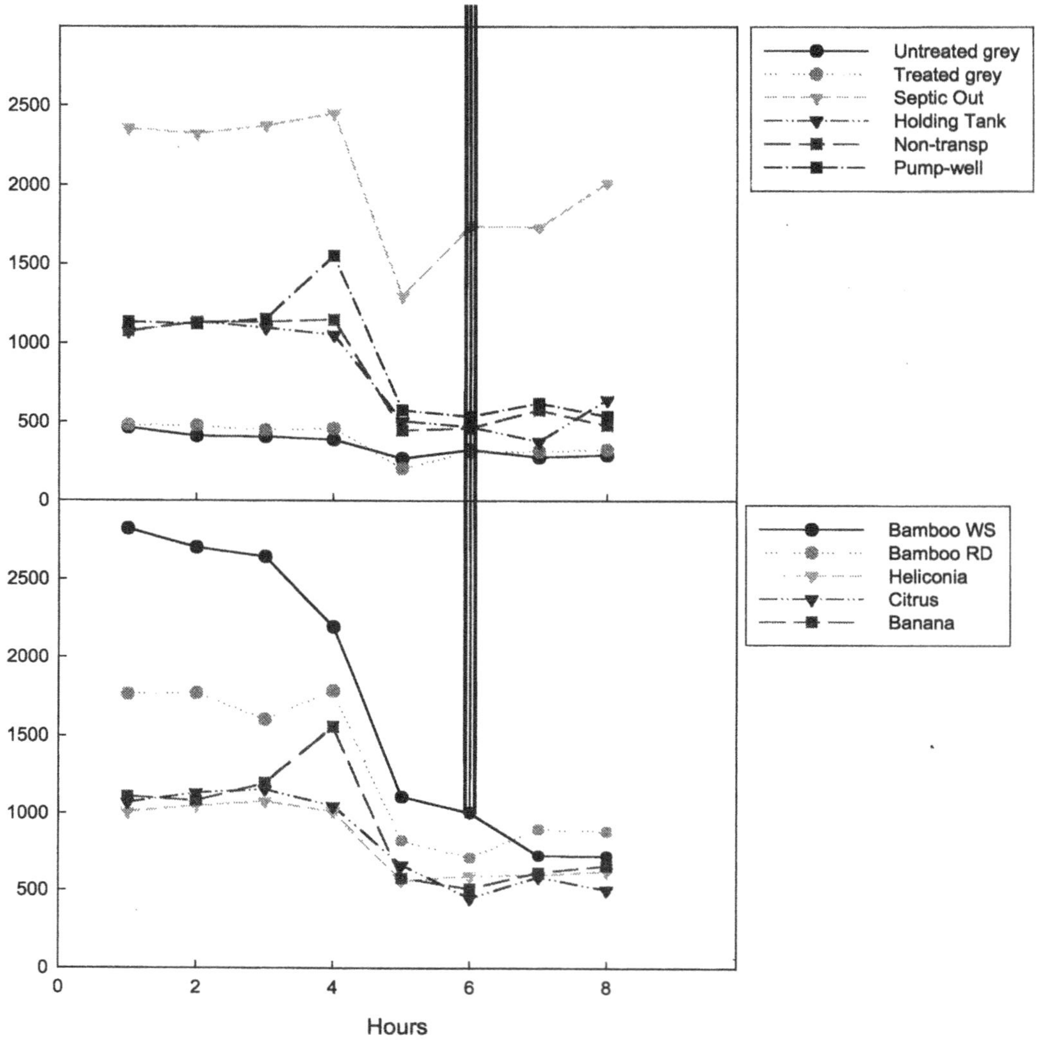

Figure 11.2.4: The effluent EC concentration at the Rockhampton site during the leaching trial.

The EC was slightly reduced in the untreated greywater and the treated greywater during the course of the trial. The EC of the reticulated town-water added to the KEWT system was 460 μS. This was close to the EC value of the greywater prior to the leaching trial; so no great change was expected. Between the 4th and 5th hour the septic outflow EC concentration decreased by about 1250 μS. The salinity in the septic discharge increased between the 5th and 8th hours until it was approximately 80% of the pre-leaching trial concentration. The EC concentrations of the effluent in the holding tank, pump-well, and non-transpired effluent all decreased by at least 50% between the 4th and 5th hour. There was no marked increase in EC concentration in these sample points between the 5th and 8th hour. The salinity in the selected KEWT channels decreased dramatically between the 4th and 5th hour. All KEWT channels had approximately 50% decrease in EC concentrations after the reticulated town-water was added and salinity did not increase between the 5th and 8th hours. The 2000 L of effluent removed from the pump-well between the 4th and 5th is thought to have contained the salts that were removed from the KEWT system. A sample of the removed effluent had an EC concentration of 2437 μS.

The major potential limiting factors in the KEWT channel soil for the Sapphire site are represented in Table 11.2.1.

Table 11.2.1: Salinity, sodicity, and CEC of KEWT system soil at the Sapphire site

Soil Sample	Sodium meq/100g	Chloride mg/kg	EC dS/m	EC (se) dS/m	Cation Ex. Cap. meq/100g	Sodium% of Cations (ESP)
Original Soil 2000	0.55	45	0.13	1.5	7.45	7.42
Citrus 2001	8.62	2150	1.44	10.6	15.54	55.47
Citrus 2002	10.30	2500	1.64	12.2	17.18	59.97
Melaleuca nervosa 2001	9.64	2200	1.51	11.2	16.39	58.80
Melaleuca nervosa 2002	10.99	2650	1.79	13.2	18.04	60.92

The original soil was suitable for the application of effluent with relatively low salinity concentrations and low ESP values. An application of a mulch/humus through the original soil would have increased the CEC and produced an even more sustainable soil for effluent application. The sodium and chloride ions, as well as the EC (se), increased dramatically at this site. There was no substantial difference between the various KEWT channels to these increases. The CEC of the soil approximately doubled, but during the same-period the ESP increased by about eight-fold. The quality of effluent added to this soil was at unsustainable and the sodium and chloride ions present in the wastewater were at concentrations that would adversely affect the soil structure.

The salinity, sodicity, and CEC values of the soil in the selected KEWT channels at the Rockhampton site are presented in Table 11.2.2.

Soil Sample	Sodium meq/100g	Chloride mg/kg	EC dS/m	EC (se) dS/m	Cation Ex. Cap. meq/100g	Sodium% of Cations (ESP)
Original Soil 1998	1.99	145	0.21	2.2	8.60	23.13
Bamboo WS 2001	2.10	175	0.28	3.2	12.90	16.85
Bamboo WS 2003	2.28	205	0.35	3.6	18.23	12.48
Bamboo RD 2001	2.05	155	0.35	3.5	15.99	8.95
Bamboo RD 2003	1.80	185	0.32	3.3	25.80	6.99
Tropical Fruits 2001	2.55	210	0.45	4.4	22.55	12.46
Tropical Fruits 2003	2.76	380	0.51	5.2	26.95	10.26
Banana 2001	2.30	125	0.21	2.1	15.68	19.84
Banana 2003	3.04	100	0.24	2.5	21.16	14.35

The original soil would not be considered ideal for the application of effluent, due to a relatively low CEC value and a high ESP percentage. However the wastewater at this site contained substantial quantities of calcium carbonates in the form of waste concrete and lime. It can be seen in the soil samples taken from the selected KEWT channels that the CEC value increased and the ESP percentage decreased, this is supported by the nutrient data which showed an increase in the calcium and magnesium ions at this site, data not presented (refer to Kele 2005). Total sodium ions in the soil at the site actually increased but other cations accumulated at a quicker rate. A slight decrease in sodium ions did occur in bamboo RD channel; the reason for this was not known. Chloride ions increased in all of the selected KEWT channels except the banana. Investigations have shown that banana plants use a small amount of chloride ions as a nutrient (Hopkins 1999). In all of the selected KEWT channels the EC (se) increased in relation to the original soil. The tropical fruit channel had the greatest accumulation of chloride ions and salinity. Soil analysis of the Rockhampton KEWT channels after the salinity leaching trial showed that the chloride ion and salinity concentrations could be reduced (Kele 2005).

Discussion

All three of the KEWT systems discussed in this paper had standard wastewater analysis conducted which showed that on average they treated water to a minimum Queensland class B recycled water standard, with a mean BOD_5 <15 mg/L, TN <20 mg/L, TP <10 mg/L, TSS <15 mg/L and faecal coliforms below <1000 cfu's at Rockhampton and Sapphire and <2 cfu's at the Gem Air caravan park, which has an add-on disinfection unit (Kele 2005).

However it can be clearly seen from Figure 11.2.1 and Table 11.2.1 that the Sapphire site had unsustainable amounts of salts within the recirculated effluent and the soil, and that the CEC and ESP of the soil were at harmful concentrations. All the plants in the irrigation area at the Sapphire site died due to the phytotoxic effects of these limiting factors. The source of the salinity and the sodicity was traced to the effluent from the chemical toilet dump-point. The chemicals used to treat and disinfect the wastewater were high in salts, sodium, formaldehyde, and acids. The Rockhampton site showed in Figure 11.2.3 and Table 11.2.2 that while salts accumulated within the recirculated effluent and within the KEWT channel soil, the CEC and ESP values actually improved over time. This was due to the characteristics of the wastewater from the concrete factory at which the site was based. The wastewater contained relatively high concentrations of calcium and magnesium ions from the calcium carbonates present in both the concrete and lime powder used at the site. So while the concentrations of sodium ions did rise, the rate of increase for the calcium and magnesium ions was proportionally higher resulting in improved soil health for CEC and ESP. The leaching trial demonstrated in Figure 11.2.4 showed that the accumulated salts within the recirculated effluent and the KEWT channel soil could be leached out of the system with a low-salinity water flush of the system, thus increasing the sustainable life of the treatment system. The Gem-Air site showed in Figure 11.2.2 that the salinity within the KEWT system effluent did not accumulate to the same extent when a quantity of effluent was diverted out of the treatment system and used for a separate recycling purpose.

The Sapphire site has been redesigned and now incorporates an acid trap, an additional specialised septic tank with reagents able to degrade formaldehyde, and KEWT channels containing a red zeolite that binds sodium and releases calcium and magnesium ions, as well as the majority of the soil being replaced with coarse sand, a medium largely unaffected by the adverse impacts of a high ESP. The plants in the Sapphire KEWT channels will be salt tolerant and a regular freshwater flush will be conducted as part of a stipulated maintenance schedule. This redesigned site was installed in June 2006, and is likely to treat the wastewater produced at this site in a sustainable manner. The characteristics of the wastewater and the associated limiting factors at this site necessitated the changes to the treatment system at this site.

Conclusion

The Rockhampton and Sapphire sites used the same treatment technology, were of a similar size and dealt with comparable volumes of wastewater. The characteristics of the wastewater treated by each system were significantly different and markedly altered respective sustainable timeframes for the land application area for the treated effluent. The closed recirculatory system allowed for the potential limiting factors to accumulate at a faster rate; however these could be controlled to an extent through management practices, such as a freshwater flush. The open recirculating system will most likely have a greater sustainable operating life. The impact of potential limiting factors on the sustainable treatment life of on-site systems should be further investigated and addressed in the relevant legislation.

Acknowledgements

The authors would like to recognise the efforts of the AWTT program, QWRS, NHT, Donna Hobbs, Robert Lowry, Charmain Elder, Tom Osborne, Billy Sinclair, Graham Doyle, Pam McKennariey, and Gavin and Kathy Kele.

References

AS/NZS (2000). Australia/New Zealand Standard On-site Domestic-Wastewater Management (AS/NZS 1547:2000). Australia/New Zealand, Joint Australian/New Zealand Standard: 1-170.

Beal, C., E. A. Gardner, *et al.* (2005). "Septic Absorption Trenches: Are They Sustainable?" *Water* **32**(1): 22-26.

Bond, W. J. (1998). "Effluent Irrigation – an Environmental Challenge for Soil Science." *Australian Journal of Soil Research* **36**: 543-555.

Gardner, T. (2005). *Keynote: Are On-site Systems Environmentally Sustainable? Can Monitoring Provide the Answer?* On-site '05, University of New England, Armidale, Lanfax Laboratories.

Graaff, R. v. d. and R. A. Patterson (2001). *Explaining the Mysteries of Salinity, Sodicity, SAR and ESP in On-site Practice.* On-site '01, University of New England Armidale, Lanfax Laboratories.

Han, F. X., W. L. Kingery, *et al.* (2000). "Accumulation of Heavy Metals in a Long-Term Poultry Waste-Amended Soil." *Soil Science* **165**(3): 260-268.

Hopkins, W. G. (1999). *Introduction to Plant Physiology*. New York, John Wiley & Sons Inc.

Incitec (2003). Methods of Soil Analysis. Brisbane, Incitec Fertilizers Ltd.

Kele, B. (2005). Onsite Wastewater Treatment and Reuse Using Recirculatory Evapotranspiration Channels In Regional Queensland. *PSG - Master of Applied Science.* Rockhampton, CQU: 1-338.

Patterson, R. A. (1997). *Soils and Effects on Effluent. What do we measure?* Proceedings Production and Environmental Monitoring Workshop, Armidale, University of New England.

Siebe, C. (1995). "Heavy Metal Availability to Plants in Soils Irrigated with Wastewater from Mexico City." *Water, Science and Technology* **32**(12): 29-34.

11.3 Worldwide Applications of Wastewater Gardens and Ecoscaping: Decentralised Systems which Transform Sewage from Problem to Productive, Sustainable Resource

Mark Nelson[1,2,3], Robyn Tredwell[1,2], Andrzej Czech[4], Made Suraja[5] and Florence Cattin[6]

[1]Institute of Ecotechnics, London, U.K. info@ecotechnics.edu

[2]Birdwood Downs Company, Derby, West Australia info@birdwooddowns.com

[3]Wastewater Gardens International, Santa Fe, New Mexico, USA email: nelson@biospheres.com website: www.wastewatergardens.com

[4]Carpathian Heritage Society, Krakow, Poland email: ac@nepcon.pl

[5]IDEP Foundation, Bali, Indonesia wwg@idepfoundation.org; www.idepfoundation.org

[6]Wastewater Gardens (WWG) Espana, Vejer de la Frontera, Spain;fc@internationalsolutions.org

Abstract

The Wastewater Garden system arose from the experience with water treatment and reuse in Biosphere 2, a closed ecological system facility supporting eight people, their technical/research laboratories, habitat and agricultural/food production and a variety of natural ecosystems in a virtually air-tight "mini-world". That initial constructed wetland, for over two years successfully treated and recycled the wastewater from humans, domestic animals and laboratory/workshops and sent its remaining nutrients and freshwater back into supply for the system's intensive agriculture irrigation system. This research experience inspired the development of high biodiversity "Wastewater Gardens" (www.wastewatergardens.com). In the years since initial research and applications along the Yucatan Peninsula in Mexico, the technology has evolved and been implemented in Belize, the Bahamas, Spain, France, the high desert of New Mexico in the United States, Poland (near Krakow and in the Carpathian Mountains), the Philippines, and in Bali, Aceh and Sulawesi in Indonesia, and in the Kimberley region of West Australia at remote rural properties and on Aboriginal communities. The high economic and ecological costs of centralized sewage treatment – pump stations, high-tech treatment plants with machinery and use of chemicals, ending with "disposal" rather than reuse of the water is in contrast to the benefits which can be realized from intelligent design of small, decentralised and on-site treatment and reuse systems. We have now added to the original concept of Wastewater Gardens – subsurface flow wetlands planted with high diversity and use of commercially valuable plants – the concept that total reuse of the water should be the objective. Treated water from the constructed wetland is used for subsoil irrigation of "ecoscapes" adapted to the landscape and climate, affording more uptake of remaining nutrients and reuse of the freshwater. These ecoscapes can use native and "bushtucker" plants; or to grow shrubs and trees which can provide fruits, timber, fibre, and medicinal plants.

Key Words

Constructed wetland, biodiversity, recycle, natural systems, subsurface flow, wastewater treatment, greening, water conservation, bioremediation, on-site sewage systems, wastewater gardens.

Introduction

The connection between wastewater and overall sustainability is far simpler to grasp in a small constructed mini-world, such as Biosphere 2 (Allen, 1991; Nelson *et al*, 1993). There it is clear that the production of "wastewater" by human activities – from crop-production, raising of livestock, laboratories and workshops, kitchens, showers, laundries and toilets – must be recycled safely back into the overall water cycle. In a materially-closed system, there is only a finite quantity of materials. Recycling is essential to keep resources available for the maintenance and evolution of life on this planet.

The constructed wetlands were a crucial part of the treatment and recycle process in Biosphere 2. Two systems were designed – one to treat all the human residential wastewater and a second to treat urine/wash down water from the domestic animals and wastewater from workshops and laboratories inside the facility. The two constructed wetlands totaled 41 m^2 of surface area with emergent and floating plants and produced a total of 720 kg, dry weight, of emergent vegetation and 493 kg, dry weight, of floating vegetation during the two-year experiment. Analysis for BOD indicated reduction was >75% with hydraulic retention times of around four days in the holding tanks and three days in the wetland treatment system (Nelson *et al*, 1999, Nelson, 1998).

Table 11.3.1: Vascular plants in the Biosphere 2 wetland wastewater recycling system during the two year closure experiment, 1991-1993 (Nelson et al, 1999).

Scientific Name	Common Name
Azolla caroliniana	Mosquito fern
Canna edulis	Canna
Canna flacida	Golden canna
Canna indica	Indian shot
Eichornia crassipes	Water hyacinth
Ipomea aquatica	Water spinach
Lemna minor	Duckweed
Pistia stratoites	Water lettuce
Phragmites australis	Common reed
Sagittaria falcate	Wapato
Sagittaria montevidensis	Giant arrowhead
Scirpus californicus	Bullrush
Spirodela polyrhiza	Duckweed
Wolffia sp.	Water meal

The wetland system was housed in several fiberglass tanks and submersible pumps maintained water recirculation between tanks. Loading to the system was on a batch basis after the primary settling tanks became full. Fourteen plant species composed the primary autotrophic level in the wetland system (Table 11.3.1). The constructed wetland system supported floating (aquatic) and emergent (rooted) wetland species. The constructed wetlands increased biodiversity within the facility and helped complete water and nutrient cycles (Nelson, 1997).

Wastewater Approaches for Remote Communities and Developing Countries

Nutrient recycling is just as central to the challenge of transforming human economic activities in Earth's vaster biospheric life support system to a sustainable basis. Wastewater treatment should also accomplish the return of nutrients and water to productive use. An important development of the past few decades has been the use of natural and constructed wetlands for the treatment of domestic sewage and industrial wastewater and the recognition of the need for water conservation and greywater recycle/irrigation.

After the Biosphere 2 experience, the lead author (MN) worked with H.T. Odum and the Center for Wetlands at the University of Florida to further the development of appropriate ecologically-based solutions for treating and reusing residential sewage especially for small, decentralised systems in remote or developing regions. Sewage treatment systems must be low-tech, low maintenance and minimal in their energy requirements to be affordable and easy to maintain in such applications, attributes which constructed wetland systems exemplify. Natural and constructed wetlands rely on solar insolation as a main driving energy, and warmer climates improve treatment rates (Kadlec and Knight, 1996). Therefore, wetland treatment systems operate more effectively in warm, high sunlight Mediterranean, arid and tropical regions. In addition, wastewater interface ecosystems benefit from the higher species diversity found in such areas since diversity at the biotic and metabolic level increases the buffering capacity of ecosystems (Mitsch and Jorgensen, 1991). Allowing self-organization of plant, animal and microbial biota to develop cooperative mechanisms may develop better adapted ecosystems to handle pollution and toxicity (Odum, 1991).

Previous studies of subsurface flow wetlands for sewage treatment have demonstrated their advantages in situations of small on-site sewage loading, in areas where land is scarce, and in situations where avoidance of malodour and mosquito-breeding are important. A well-designed subsurface flow wetland also can provide highly effective sewage treatment. The advantages of constructed wetlands are that because they rely on more natural methods, they are often less expensive to build and far less expensive to operate than conventional sewage treatment plants. They can also produce a standard of treatment comparable to advanced wastewater treatment (Reed *et al*, 1995). Subsurface wetlands use little or no electricity and technology and require little technical supervision once installed. Although more expensive per unit area than free-surface wetlands, because of the cost of impermeable liner and gravel media, subsurface flow constructed wetlands have more intense treatment action, reducing required area significantly (Green and Upton, 1992; Steiner *et al*, 1992; Cooper, 1992).

Applications in the Yucatan Peninsula, Southern Mexico

Commencing in 1996, several dozen Wastewater Garden, subsurface flow constructed wetlands were installed for houses, condominiums, restaurants and small hotels south of Cancun, Mexico in the Yucatan Peninsula coastal region. This area is highly karstic with soils very permeable to the flow of wastewater which pollutes the shallow water tables and adversely affects the coral reef adjacent to the coastline. Experiments were commenced with use of a high biodiversity of plants, both native and decorative/valuable ones already in the region. Most of these systems were completely gravity-flow, but for some installations, the constructed wetlands were designed as raised garden boxes and submersible pumps raised effluent from the septic tanks to the systems. Table 11.3.2 presents the water quality data over a two year period from the first two Wastewater Gardens® (Nelson, 1998a, 1998b). Discharge water from the systems was sent to subsurface leachdrains and in one case studied, to a nearby mangrove area whose self-created high-organic peat soils offered a very appropriate final biofilter for remaining nutrients in the treated wastewater. However, with the passage of time, many systems suffered from poor maintenance such that taller vegetation was not pruned back and was allowed to shade out understory plants, resulting in a loss of effective wetland biomass and diversity. It is also suspected that system design, with short distances to discharge pipes may have resulted in short-circuiting. Subsequent testing showed COD reduction had fallen to 65-70% and nutrient reductions had declined as well. Coliform bacteria reduction still appears to be 99+% (CEA, unpublished data).

Table 11.3.2: Removal efficiency and loading of Wastewater Garden subsurface flow systems Studied at two initial locations in Mexico (1996-1998) (Nelson, 1998a, Nelson, 1998b).

Parameter	In (Septic tank discharge) mg/l	Out (discharge WWG) mg/l	Removal %	Loading kg/ha/d
BOD_5	145	17.6	87.9	32.1
Total Phosphorus	8.05	1.9	76.4	1.7
Total Nitrogen	47.6	10.0	79	10.3
Total Suspended Solids	69.9	38.9	44.4	15.1

	In	Out	Removal %
Coliform bacteria	4.9 x 106	2.2 x 103	99.8%

These lessons have led to several changes in subsequent design approach:

1. The realization that the goal should be as close to total productive use of water and nutrients – not simply the level of removal in the constructed wetland. This has changed our conception from the old paradigm of "final disposal, e.g. leachdrains" to incorporating greywater recycle/irrigation where possible and designing a robust subsoil irrigation of the treated effluent from the Wastewater Gardens and dealing with total recycling/reuse of wastewater not simply "treatment".

2. Length to width ratio needs to be at least 4:1 to lessen chance of short-circuiting and use of internal baffles in the wetlands where possible.
3. Maintenance manuals were upgraded to emphasize the need for attention to the systems, especially filters in septic tanks, pruning for maintaining good numbers and diversity of vegetation, especially that of deeper-rooted wetland plants, and removal of any organic matter which begins to build up on the gravel surface.
4. For situations which demand higher initial treatment performance, use of vertical flow design which results in better aeration and thus reduction of N and BOD. These systems generally require use of dosing siphons or pumps to ensure batch-loading and are thus more technically-demanding and reliant on electricity.

In the years since, Wastewater Gardens International has operated both through doing design consultation and implementation world-wide on a project basis, and has worked with regional representatives to carry forth applications in selected regions. These include: Indonesia, Poland including the Carpathian Mountains, Spain/Portugal and North Africa, and in the Kimberley region of West Australia.

Wastewater Gardens in Indonesia

Wastewater Gardens® came to Indonesia in 1999 to fill a gap in available technologies on the island by providing an effective wastewater treatment solution for resorts, hotels, businesses and communities at a price affordable in terms of the local economy and manageable in terms of local engineering and construction expertise. The UNDP projects that on current rates of use and existing policies for the expansion of tourist facilities, water supply for Bali would be in deficit within a decade, and demand would reach four times potential supply by the middle of the next century (Warren, 2000). Also the vast majority of Bali's and Indonesia's wastewater goes completely untreated, fouling rivers, groundwater and adding to environmental damage. The consequences for human health are also enormous. After the success of this project, IDEP Foundation, a local environmental education organization working to train Indonesians in environmentally friendly solutions (www.idepfoundation.org), set up a Wastewater Garden division, working with Nelson to implement constructed wetlands. The early years of this partnership saw 15 more WWGs installed at hotels and resorts across Indonesia including one at the Tirtagangga Water Palace, an icon of Balinese culture funded by the Seacology Foundation and Livingry Foundation. Also of note was a village bamboo treatment centre using mangroves in the Wastewater Garden as an educational demonstration of the importance of mangrove conservation (working with the Mangrove Action Project) in a fishing/farming village near Menado, Sulawesi. In recent years, work includes training of local designers/builders and research into alternative materials and applications for the technology. Notably, the system is being implemented in post-tsunami reconstruction efforts in Aceh, Sumatra, Indonesia.

Wastewater Gardens Projects in Poland – Promotion of Decentralised Wastewater Treatment Systems

The first Wastewater Gardens were installed in the Carpathian Mountains in the year 2000 in cooperation with a local NGO called Carpathian Heritage Society (www.carpathians.pl/english.html). A WWG system was built for Research Station of the Jagiellonian University in Krempna in the Magurski National Park which purifies wastewater from 20 people. Treated wastewater feeds adjoining orchard.
Since year 2002, Twelve Wastewater Garden systems have been built in Poland – mostly for private houses or businesses. These Wastewater Gardens™ system comply with Polish environmental regulations, European standards, United States Environmental Protection Agency (EPA), and German ATV. It is confirmed by results from the pilot Wastewater Garden in Krempna (Table 11.3.3). That discharge water goes out to subsurface irrigation of a fruit orchard.

Table 11.3.3: Results from the pilot Wastewater Gardens system in Krempna, Poland in the Magurski National Park. *ccording to Polish environmental regulations.

Parameter	Influent concentration	Effluent concentration	Percent reduction	Limit value*
BOD_5	55.0 mg O_2/l	11.0 mg O_2/l	80%	40 mg O_2/l
COD	88.0 mg O_2/l	32.0 mg O_2/l	64%	150 mg O_2/l
SS	74.5 mg/l	49.5 mg/l	34%	50 mg/l
Total N	73.7 mg N/l	24.6 mg N/l	67%	30 mg N/l
Total P	7.2 mg P/l	2.0 mg P/l	72%	5 mg P/l

Spain/Portugal and North Africa (Morocco/Algeria)

In Spain and Portugal, working through a new Division of Wastewater Gardens International, we have been approached to implement projects for homes, hotels and small municipalities as additional treatment to existing sewage treatment facilities. Also in planning are several demonstration projects in North Africa: a first one was implanted in Temacine, southern Algeria in early 2007 serving 100 people.

The Bahamas: a 2-cell, 80 m^2 system followed by subsoil irrigation was implemented at the Cape Eleuthera Island School, a school which focuses on environmental and science hands-on education. The system serves all the school and residential buildings.

Puerto Rico: a 52 m^2 WWG serves all the grey and blackwater at the homestead buildings of the Las Casas de la Selva rainforest enrichment and forestry project (www.eyeontherainforest.org) near Patillas, Puerto Rico. The Department of Natural Resources funded the project as a demonstration ecological approach to sewage recycling and to protect mountain streams which feed water reservoirs for local populations. Water quality testing showed over 90% reductions of BOD, TSS and total N, total P reduced by 83% and a 99.9% reduction in coliform bacteria.

Belize: the Kanatik Reef and Jungle Resort installed two Wastewater Gardens to treat their restaurant, laundry and guest rooms, one of 200 m^2 and the other 120 m^2 to handle a total of 90-100 guests and staff.

France: a Wastewater Garden was installed at the Les Marronniers conference centre and ecological farm near Aix-en-Provence which treats all the conference centre kitchen and residential sewage.

The Philippines: Wastewater Garden systems were installed at the Children's Village project of the ABS-CBN foundation outside of Manila.

Applications in the Kimberley Region of West Australia

In 2000, the West Australian Department of Health approved the commencement of some pilot projects using the Wastewater Garden approach in the Kimberley of northwest Australia, since remote towns and indigenous communities need practical, decentralized solutions to sewage treatment and reuse. Working with Birdwood Downs Company (www.birdwooddowns.com) which has been implementing environmental upgrade and integration of sustainable approaches in the region (Allen *et al*, 1984; Nelson, 1985), five projects have been implemented to date:

1. Birdwood Downs homestead kitchen/dining room/library/toilet and shower: an 8 m^2 Wastewater Garden was sunk 0.5m into the ground to permit gravity-flow from an existing septic tank. The system handles half the wastewater from a resident and tourist population from 5-12. The remaining wastewater is used for subsoil irrigation of a banana and fruit tree patch near the ablution block. The Wastewater Garden system has operated without problem for over 7 years and was monitored for water quality improvement (Table 11.3.4). The garden supports heliconia (Bird of Paradise), canna lilies, pandanus palm, coconut palm, plantain, 2 types of elephant ear, papyrus and oleander. Treated effluent irrigates other trees in the back of the homestead

Table 11.3.4: Summary of water quality results from Wastewater Garden system at Birdwood Downs homestead, Derby, West Australia. Analyses were conducted at MPL Laboratories, Perth, an accredited testing facility. If evapotranspiration averages 20% in above system, total reduction of BOD is 96%, reduction of TSS is 96%, reduction of N is 59%, reduction of P is 43%, and reduction in coliform bacteria is 99%.

Parameter	BOD-5 (Biochemical Oxygen Demand) mg/l	Total Suspended Solids (TSS) mg/l	Total Nitrogen mg/l	Total Phosphorus mg/l	Total Coliforms Cfu/100 ml
Average in Septic Tank	241	338	52.5	10.3	6,285,000
Average Wastewater Garden discharge	12	16.5	27	7.3	116,000
Percent Reduction	95%	95%	48%	30%	98.2%

2. Emu Creek (Gulgagulganeng) community, Kununurra, West Australia

Several of the houses at Emu Creek community are on very low-lying soils, adjacent to natural wetlands, and since the community was moved to this location, there have been problems with failure of leachdrains, sewage on the ground and house plumbing backing up. For a small community (30-50 residents) centralised solutions, such as sewage lagoons, are very expensive and would require daily power and technical maintenance. The solution implemented with community participation in the planning was three Wastewater Garden systems, constructed in June 2002, to serve the various houses on the community. To minimize dependence on electricity and machinery, the two systems serving the lower three houses were sunk into the ground (in one case reaching groundwater at 0.7m below surface in June) so that septic tanks can discharge through pipelines by gravity-flow into the Wastewater Gardens. For these two systems, dual submersible pumps were installed in the control box after the WWG, so that during the wet season if the ground gets saturated, the gravity-flow leachdrain can be turned on, and the pumps will discharge subsoil to an area of higher elevation sandy soils which never get saturated. The third system serves a house on the higher elevation side of the community and only required a gravity-flow system. Two years of water quality tests after the systems were installed (Table 11.3.5) show relatively high levels of treatment are being achieved.

Table 11.3.5: Summary of water quality results at Wastewater Gardens at Emu Creek community, Kununurra. West Australia. Analyses were conducted at MPL Laboratories, Perth, an accredited testing facility. If evapotranspiration is 20% for above system, then wetland reduction of BOD is 91%, reduction of TSS is 92%, reduction of nitrogen is 77% and reduction of P is 65%. Distance to laboratory prevented measurement of coliform bacteria.

Parameter	BOD-5 (Biochemical Oxygen Demand) mg/l	Total Suspended Solids (TSS) mg/l	Total Nitrogen mg/l	Total Phosphorus mg/l
Average in Septic Tank	214	99	228	18
Average Wastewater Garden discharge	23	10	66	7.8
Percent Reduction	89%	90%	73%	58%

3. Coco-Eco Bed and Breakfast, Coconut Well, Broome, West Australia

The Coco-Eco B&B, Coconut Well, Broome is a ecologically-designed facility run on solar electric power which wanted to set a high standard for treatment and recycle of wastewater by using them to create lush, tropical gardens. Their 16m^2 Wastewater Garden is fit into the courtyard centre of the facility and remaining nutrients are used in their subsoil irrigation leachdrain area to support additional decorative flowering shrubs and productive crops like banana. Built in February, 2004, the facility was awarded the 2005 Laminex Group Single Residential Award from the Royal Architectural Institute of Australia in recognition of its ecological design and features.

4. Joy Springs (Eight Mile) community, Fitzroy Crossing, West Australia
At the Joy Springs (Eight Mile) community near Fitzroy Crossing, West Australia four of the houses on the low-elevation side experience leachdrain failure and plumbing backup during wet seasons when soils were flooded. As a solution, the houses were retrofitted in June/July 2004 with individual $12m^2$ Wastewater Gardens and gravity leachdrains/inverted leachdrains. The inverted leachdrains were planted with a mix of native trees and shrubs (especially valuable bush-tucker and medicinal plants) and the area is outfitted with a drip irrigation reticulation system on automated timer to provide irrigation for these backyard gardens as a backup source of water.

5. Pandanus Park community, Derby, West Australia
In 2007, with funding from the Australian Community Water Grants and the Department of Housing and Works, WA Government, six houses and a childcare centre at the Pandanus Park aboriginal community near the Fitzroy River in Derby, W.A. were connected to three Wastewater Garden systems and their outflow was used for final subsoil irrigation in native plant and shrub ecoscape areas. These systems are all gravity-flow systems without the need for pumps.

Conclusions

The world is experiencing increasingly the importance and scarcity of our sources of fresh, clean water – and the necessity to improve human health and protection of our environment, both of which are threatened by contamination by human wastewater.
In this context, new approaches which can replace or augment centralized high-tech wastewater treatment are crucial. Those in developing countries, and in remote and small communities of developed countries, can neither afford the considerable infrastructure and machinery upkeep/maintenance requirements of centralized sewage treatment, nor will they benefit as much as by intelligent on-site, decentralised systems. The advent of a class of more ecologically-based approaches, such as using wastewater for greening, either through constructed wetlands and/or greywater recycle/irrigation creating locally-adapted "ecoscapes" offers some new alternatives which may be part of the solution to the looming 21st century water crisis of Australia and the world. In our planetary biosphere, our ultimate life support system, water conservation and wastewater reuse are key elements for both human and biospheric health.

References

Allen, J.P., 1991. *Biosphere 2: The Human Experiment.* Penguin Books, NY.

Allen, J.P., Nelson, M., and Snyder, T.P., 1984. Institute of Ecotechnics, *The Environmentalist* 4 (1984) 205-218, Geneva.

CEA (Centro Ecologico Akumal, Mexico), unpublished data, Kate Riley 2005.

Cooper, P.F., 1992. The Use of Reed Bed Systems to Treat Domestic Sewage: The present Situation in the United Kingdom, pp. 153-172 in *Constructed Wetlands for Water Quality Improvement*, Moshiri, G.A. (ed.), Lewis Publishers, Boca Raton, FL.

Green, M.B. and J. Upton, 1992. Reed bed treatment for small communities: U.K. experience, pp. 518-524 in *Constructed Wetlands for Water Quality Improvement*, Moshiri, G.A. (ed.), Lewis Publishers, Boca Raton, FL.

Kadlec, R.H., and Knight, R.L., 1996. Treatment Wetlands, Lewis Publishers, Boca Raton, FL.

Mitsch, W.J. and S. Jorgensen (eds.), 1991. Ecological Engineering: An introduction to ecotechnology, Wiley & Sons, NY.

Nelson, M., 1998a. Wetland systems for bioregenerative reclamation of wastewater -- from closed systems to developing countries, *Life Support and Biosphere Science* 5(3): 357-369.

Nelson, M., 1998b. *Limestone mesocosm for recycling saline wastewater in coastal Yucatan, Mexico*, Ph.D. dissertation, Dept. of Environmental Engineering Sciences, University of Florida, Gainesville, pp. 329.

Nelson, M., T. Burgess, A. Alling, N. Alvarez-Romo, W. Dempster, R. Walford, and J.P. Allen. 1993. Using a closed ecological system to study Earth's biosphere: Initial results from Biosphere 2. *BioScience* 43(4): 225-236.

Nelson, M., 1985. "Synergetic management of the savannahs", paper presented at the Second International Savannah Symposium, CSIRO, Brisbane, Queensland, Australia. Published, pp.190-192, in *The World's Savannahs: Ecology and Management*, Australian Academy of Science, (ed. J. Mott and J. Tothill) University of Queensland Press, Brisbane, Australia.

Nelson, M. and Tredwell, R., 2002. "Wastewater Gardens: Creating urban oases and greenbelts by productive use of the nutrients and water in domestic sewage", paper for *Cities as Sustainable Ecosystems (CASE) Conference*, April 2002 Environmental Technology Centre, Murdoch University, Perth, Australia.

Nelson, M., Finn, M, Wilson, C., Zabel, B., van Thillo, M., Hawes, P., and R. Fernandez, 1999. Bioregenerative recycle of wastewater in Biosphere 2 using a created wetland: two year results, Ecological *Engineering* 13: 189-197.

Nelson, M., 1997. Nutrient recycling systems of Biosphere 2: litterfall, decomposition and wastewater recycling, results from the 1991-93 closure experiment, *Life Support and Biosphere Science*, 4(3/4):145-153.

Odum, H.T., 1991. Ecological Engineering and Self-Organization, pp. 79-101, In: *Ecological Engineering: An introduction to ecotechnology* (W. Mitsch & S. Jorgensen, eds.), Wiley & Sons, NY.

Reed, S.C., Crites, R.W., and E.J. Middlebrooks, 1995. *Natural Systems for Waste Management and Treatment*, 2nd ed., McGraw-Hill, NY.

Steiner, G.R., Watson, J.T. and K.D. Choate, 1992. General design, construction and operation guidelines for small constructed wetlands wastewater treatment systems, pp. 499-507, in *Constructed Wetlands for Water Quality Improvement*, Moshiri, G.A. (ed.), Lewis Publishers, Boca Raton, FL.

Warren, Carol, 2000. The Cultural and Environmental Politics of Resort Development in Bali, Institute for Sustainability and Technology Policy, Murdoch University, West Australia http://wwwistp.murdoch.edu.au/publications/e_public/Case%20Studies_Asia/bali_2/case.htm

12 Innovative Technologies

12.1 Performance Modeling of Full-scale Desalination Processes with Highly Permeable RO Membranes

Lianfa Song, Kwee Guan Tay, Gurdev Singh

Division of Environmental Science & Engineering, National University of Singapore, 10 Kent Ridge Crescent, Singapore 119260

Abstract

In this paper, the performance of the full-scale RO process with highly permeable membranes and the governing mechanisms were carefully studied. It was found that the performance of a full-scale RO process could be controlled by two possible mechanisms, namely mass transfer rate and thermodynamic limitations. Under relatively low driving pressure, it was controlled by mass transfer rate (water flux) of the membrane. However, with the highly permeable membrane, it is possible that the performance is limited by the thermodynamic limitation, in which the osmotic pressure becomes equal to the driving pressure inside of the membrane channel. A process controlled by thermodynamic limitation is an extremely case of the hydraulic imbalance problem. When it occurs, it means part of the membranes in the processes do not contribute to permeate production. More complicated are situations in the intermediate pressure range, in which both mechanisms contribute to, but none of them can dominate, the performance of the process.

Some innovative concepts and theories on the performance of the full-scale RO processes were developed. These concepts and theories may provide better qualitative explanations for the behaviors often observed in the full-scale RO processes. A better quantitative simulations or predictions of the performance of the process were developed upon these concepts and theories. Experiments were carried out on a pilot membrane process of 6 meter membrane channel to imitate the performance of the full-scale RO under various conditions. The experimental performance data were compared with theoretical simulations and excellent agreement was obtained.

Another focus of this current study was on characterization and modeling of membrane fouling in the full-scale RO process. Colloidal fouling experiments were conducted to study the fouling potential of feed water and a new fouling indicator was proposed. The indicator can be directly used in the mathematical model to simulate fouling development in the full-scale RO processes. Model simulations showed that under certain condition (thermodynamic restriction), the recovery or average permeate flux of a full-scale RO process would maintain a constant value even membrane fouling was taking place. Experimental verification of the simulation results are currently under way. With the new developments and findings in this area, methods or protocols for optimization of full-scale processes of the highly permeable RO membranes were suggested.

Introduction

Application of reverse osmosis (RO) processes for water and wastewater treatment has been growing rapidly in recent years and it has become one of the major technologies for producing potable water in many places in the world [1,2]. However, most of our knowledge about RO membranes still remain on the small scale processes employed in laboratories. A few years ago, my research group attempted to develop a comprehensive model to simulate the full-scale RO system, which was intended for optimization of the system design and operation. One key task of the comprehensive model is to predict the permeate flux from a long RO channel (the length of the RO channel is usually 6 or 7 meter in one stage).

Basic membrane transport equation

Literature review revealed that most published studies on the full-scale RO processes predicted the average permeate flux with the basic membrane transport law [3-6]:

$$v = l(\Delta p - \Delta \pi) = \frac{(\Delta p - \Delta \pi)}{R_m} \quad (1)$$

where v is permeate flux, Δp is the driving pressure, $\Delta\pi$ is the osmotic pressure, l is permeability of the membrane, and R is the resistance of the membrane.

Although the applicability of Eq (1) to a small piece of membrane is no doubt verified with numerous experiments, the direct use of Eq (1) to the full scale RO processes is not appropriate because the osmotic pressure in the channel is no longer a constant. There is a notable increase in osmotic pressure along the membrane channel as permeate is being produced.

Cascade calculation method

It was also noted from industry practices that the accuracy of permeate flux prediction from a full-scale RO process could be improved by the so-called cascade calculation method, in which the basic membrane transport law (Eq (1)) was applied piecewise to each element in the pressure vessel. Average osmotic pressure of the feed end and the concentrate end of the element, $0.5(\Delta\pi_f + \Delta\pi_c)$, was used in the cascade calculation [1]. The osmotic pressure at the concentrate end of the membrane element was initially guessed and gradually approached to a convergent value with an iteration scheme.

The cascade calculation method usually can produce predictions with quite high accuracy for the old generation of membranes, which were operated at driving pressures that was much higher than the osmotic pressure at the concentrate end. For example, when the membrane resistance is 10^{12} Pa.s/m and salt concentration of the feed water is 5,000 mg/l, a driving pressure of about 800 psi is required for a 6-meter RO channel to reach about 50% recovery. At this time the salt concentration at the concentrate end of the RO channel is 10,000 mg/l. The osmotic pressure of the concentrate is about 100 psi, which is only one-eighth of the driving pressure. For such situations, permeate flux of the pressure vessel can be excellently predicted with the cascade calculation.

The currently used RO membranes usually have resistance that is more than one order lower in magnitude than the old generations. Consequently, the driving pressure is significantly reduced. For the above example but with a new membrane of resistance of 10^{11} Pa.s/m, a driving pressure of about 120 psi may reach 50% recovery. In this case, the net driving pressure changes from 70 psi at the feed end to 20 psi at the concentrate end. The net driving pressure and the resulted permeate flux change more drastically in the membrane channel with the more permeable RO membranes (Fig. 12.1.1). When the osmotic pressure is comparable to the driving pressure, the accuracy of the cascade calculation method deteriorates significantly and sometimes it fails to produce convergent results.

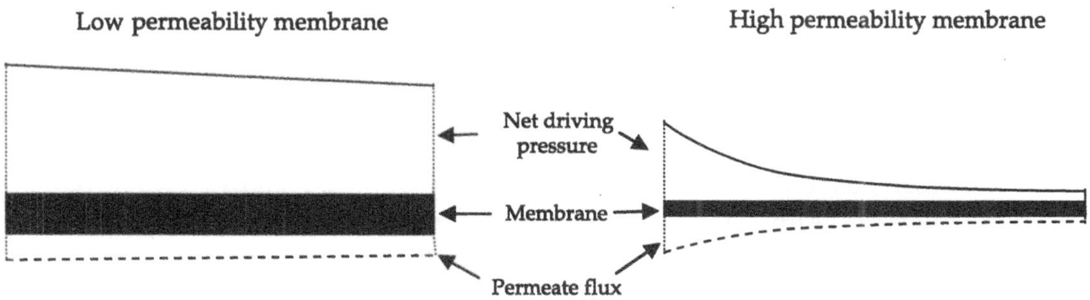

Figure 12.1.1: Variations of net driving force and permeate flux along membrane channel with low and high permeable membranes

Finite difference method

To further improve the accuracy of permeate flux calculation in the full-scale RO process, a finite difference method was developed for the membrane processes with highly permeable RO membranes. The whole channel was first divided into many small segments so that the properties in each segment can be treated as constant. In each segment, the permeate flux v, cross flow velocity u, salt concentrate c, and pressure drop Δp were determined with:

(Membrane transport law) $$v_i = \frac{\Delta p_{i-1} - \Delta\pi(c_{i-1})}{R_m} \quad (2)$$

(Mass conservation on water) $$u_i = u_{i-1} - v_{i-1}\Delta x/H \quad (3)$$

(Mass conservation on salt) $$c_i = \frac{u_{i-1}c_{i-1} - (1-r)v_i \Delta x/H}{u_i} \quad (4)$$

(Pressure loss due to friction) $$\Delta p_i = \Delta p_{i-1} - \frac{12k\eta}{H^2}u_{i-1}\Delta x \quad (5)$$

where Δx is the length of the membrane segment, H is the height of the channel, r is the salt rejection rate, k is the friction coefficient, and η is the viscosity of water.

Although the principle or concept is similar, the advantages of the finite difference method over the cascade method are obvious. First, the membrane transport and conservation principles are applied locally to small segments of membranes. Because the osmotic pressure in the small piece can be treated as a constant, no iteration is needed. For the same reason, there is no overshooting as in the cascade method there iteration is used. Secondly, the finite difference method is more flexible. The total number of the segments used in calculation can be easily adjusted to fit the requirement of more permeable RO membranes. Thirdly, the finite different method is robust that will produce the result for any conditions of operation and never break down.

Therefore, the initial goal of our research was to develop the finite difference method that can produce a more accurate, more stable and robust result for the prediction of the permeate flux in the full-scale RO processes. The result of model development was published in 2002 [7].

If our investigation stopped at this conjecture, we have to confess that the phenomena have been addressed above were obvious to many people who were knowledgeable of the real-world RO processes. Although the finite difference model has its own merit in more accurate simulation of the full-scale RO processes that is required for optimization of process design and operation, fundamentally it represents only small incremental contributions to the existing knowledge pool of RO processes.

Emergence of Thermodynamic Restriction

We used the finite difference model extensively to investigate the performance of the long RO channel with different parameters under various operating conditions, including membrane resistance, driving pressure, salt concentration, and channel length. To our big surprise, the simulation results showed us that with the RO membrane of resistance Rm equal to or smaller than 1011 Pa.s/m, there were many interesting behaviors of the full-scale RO under common operating conditions, which had not been reported or discussed in the literature.

Thermodynamic restricted flux

From the day one of invention of the RO process, it was known the permeate flux would be zero when the driving pressure was equal to the osmotic pressure, i.e., $(\Delta\pi - \Delta p)=0$. However, it was not a common knowledge that a full-scale RO could be controlled by the zero net driving force under the common operating conditions employed in the practical RO processes.

Our simulating study with the finite difference model showed that the linear law between flux and driving pressure broke down even concentration polarization was completely not considered (Fig. 12.1.2). When the channel was sufficiently long, the average permeate flux decreased with the increasing length of the channel (Fig. 12.1.3). In this case, if the permeate recovery was plotted vs. channel length, it would be a constant invariant with channel length (broken line). These interesting behaviors of the full-scale RO processes prompted us to look for the underlying reason or mechanism for the constant recovery that was not affected by the length of the membrane channel.

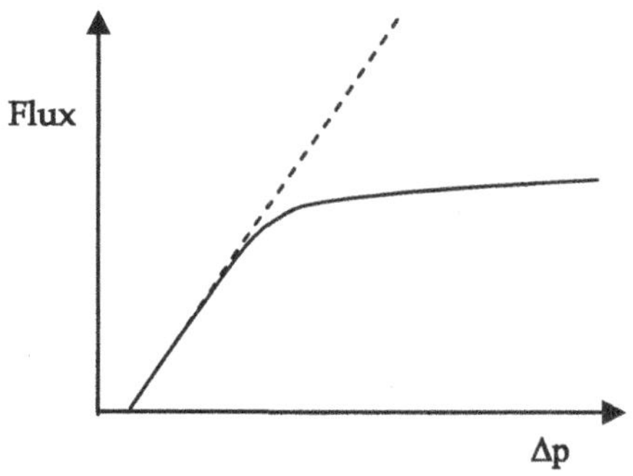

Figure 12.1.2: Non-linear relationship of v vs Δp

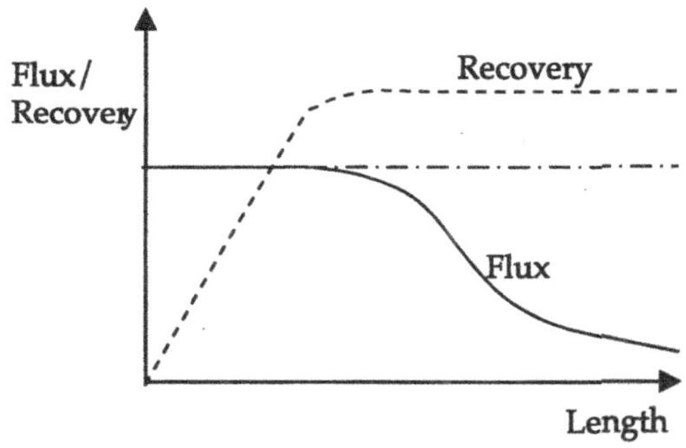

Figure 12.1.3: Average flux decreases with channel length

The finding of the non-linear dependence of the average permeate flux on the driving pressure was not planned by us. Although it was not known to us before, it must have been known for a long time to many researchers and engineers, especially to these who experienced with module testing and practical process design. It was a little surprise to me that the behavior has not been extensively reported or discussed in the literature. The behavior might be thought too obvious or too trivial to discuss. It also might be because it violated the linear relationship that has been accepted in this filed for a long time, or it could not be explained from the known principles, or for other reasons.

We believed that the above behaviors of the full-scale RO was significant and hinted another limiting mechanism for the process. It did not take us long to find out that the limitation was actually thermodynamic equilibrium. In a long membrane channel with highly permeable membrane, the equilibrium salt concentration can be reached inside the channel (Fig. 12.1.4). After the point at which the equilibrium is reached, the rest of the channel does not produce any permeate-dead region. Therefore, the average flux decreases as the dead region increases. This mechanism explains well the behavior of average permeate flux reported in Fig. 12.1.3.

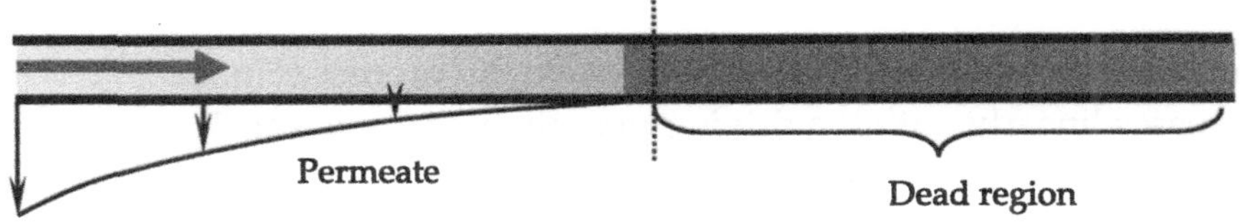

Figure 12.1.4: Substantial variation of the permeate flux along the channel and dead region

Calculation of thermodynamic restricted flux

When a process is known at thermodynamic equilibrium, there is no need to use the cascade method or the finite difference method to calculate the permeate flux or recovery. The recovery can be determined easily from thermodynamic principle, i.e.,

$$R = 1 - \frac{\Delta\pi_0}{\Delta p} \tag{6}$$

where R is the recovery and $\Delta\pi_0$ is the osmotic pressure of the feed water. With the conversion coefficient (respect ratio) Hv_0/L, the average permeate flux is determined as:

$$\bar{v} = \left(\frac{Hv_0}{L}\right)\cdot R = \frac{Hv_0}{L}\left(1 - \frac{\Delta\pi_0}{\Delta p}\right) \tag{7}$$

Eqs. (6) and (7) are important because they give us the upper limit of the performance of full-scale RO under the given salt concentration and driving pressure. The most direct use of these two equations is to evaluate the full-scale RO processes. If a process has recovery or average flux much smaller than the values calculated with these two equations, the process can be improved by prolonging the membrane channel and increasing mass transfer coefficient. If the recovery or average flux is near to the predicted value of Eqs. (6) or (7), the process efficiency cannot be further increased by the above measures.

Condition for thermodynamic equilibrium

It is demonstrated that there are two possible controlling mechanisms for the performance of the full-scale RO processes, namely mass transfer (Eq. 1) and thermodynamic equilibrium (Eq. 6 or Eq. 7). Mass transfer would be the controlling mechanism under low pressure because the low production of permeate should not cause significant change in salt concentration in the flow. The thermodynamic restriction would take over under sufficiently high pressure due to the substantially elevated salt concentration.

There was a transition pressure, ΔP_T, for each full-scale RO process (Fig. 12.1.5). The average flux increases linearly with the driving pressure when it is smaller than the transition pressure, while increases marginally when the driving pressure is higher than the transition pressure. With the knowledge of transition pressure, the appropriate equation can be chosen for the calculation of permeate flux. Further study resulted in an analytical formula of the transition pressure, which can be expressed with the basic parameters of the full-scale RO processes [10]:

$$\Delta P_T = \frac{u_0 H R_m}{L} = \frac{Q}{S}\cdot R_m \tag{8}$$

where Q is the feed flow rate to a pressure vessel and S is the total membrane area. The transition pressure can be expressed as the product of the feed flow rate per unit membrane area and the membrane resistance. The linear dependence of the transition pressure on the membrane resistance indicates the enormous effect of the improvement of RO membrane permeability on the reduction of the transition pressure. To our surprise, the transition pressure is not related to the salt concentration in the feed.

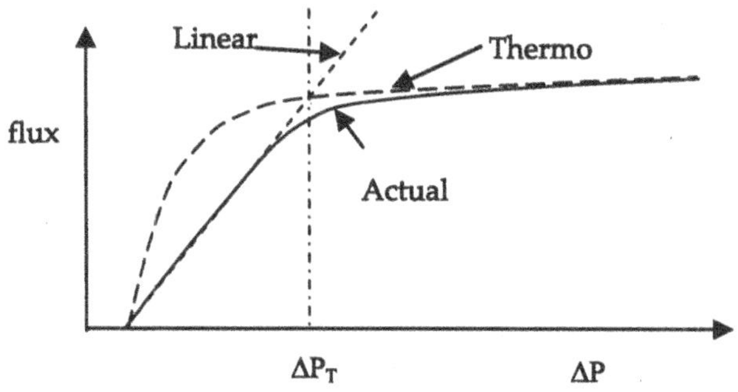

Figure 12.1.5: Transition pressure and two distinguished regimes

Working pressure

The knowledge of the characteristic pressure can help to determine the working pressure of membrane process, which was not able to be done before. Fig. 12.1.9 is a schematic description of the guideline. For any RO process, the recovery-pressure (or average flux-pressure) curve can be developed with the method presented in this paper. The typical curve has a turning point (at pressure P_1 in Fig. 12.1.9) that divides the pressure domain into two regions. In the region of lower pressure, recovery (or average flux) basically increases with pressure linearly. In the region of higher pressure, the recovery (or average flux) only increases marginally with pressure. It is obviously that P_1 is the optimal working pressure in the sense of recovery (or flux) increment per unit pressure. It can be seen that when pressure increase significantly from P_1 to P_2, the recovery increment $R_2 - R_1$ is rather small.

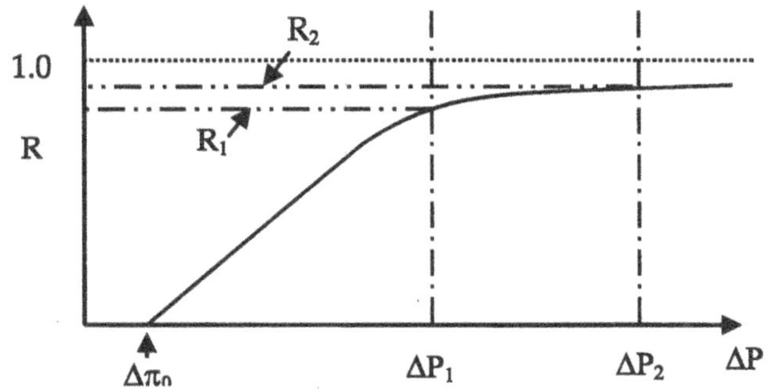

Figure 12.1.9: Schematic for determination of working pressure for a RO membrane process.

It is worth to point out that the turning point is obvious for low salt concentration. As the salt concentration increases, the turning point may become less obvious and the recovery (or flux) shows a strong nonlinear dependence over the entire pressure domain and the rate of flux increase gradually decreases with pressure. However, the determination of the working pressure for this case is less critical.

Evidences of thermodynamic restriction in literature

There were plenty of indirect evidences of the thermodynamic restriction in the literature. The first evidence was the significant reduction of the driving pressure in the practical RO processes in recent years. RO membrane used to be operated at very high driving pressure (> 500 psi) but has been reduced gradually to around 100 psi for brackish water. One obvious reason to operate RO at much lower pressure is that the increment of permeate flux is only marginal when the RO process hits the wall of thermodynamic restriction. Another piece of evidence was the reports of the imbalanced permeate production rate along the membrane channel [8, 9]. It was reported that the flux decreased rapidly along the membrane channel and the last a few membrane elements in a long pressure vessel do not contribute much to the permeate production.

Effective Fouling Characterization

It demonstrated that when a full-scale RO was operated under thermodynamic restriction, the average permeate flux was NOT affected by change in membrane resistance. A new method was proposed for a better fouling characterization that is valid for membrane fouling under any conditions.

Water fouling potential

The fouling strength of the feed water is usually quantified in numerical values with some parameters, such as SDI, MFI. It is desirable the parameter used to define water fouling potential can be used to predict the development of fouling in full-scale RO processes. For this purpose, water fouling potential, k, was defined by Eq. (9):

$$k = \frac{R_t - R_0}{\int_0^t v(\tau)d\tau} \tag{9}$$

where t is the filtration time, R_t and R_0 are the nominal membrane resistance at time t and 0 respectively, and τ is the dummy integration variable. The water fouling potential can be determined with a small piece of membrane in a lab filtration device in a few hours.

Modeling of fouling development in full-scale RO processes

When the feed water fouling potential is known, the membrane resistance in the whole membrane channel at any time can be calculated with Eq. (10):

$$R(x,t) = R_0 + k\int_0^t v(x,\tau)d\tau \tag{10}$$

Combined with Eqs (2)-(5), the permeate flux in the membrane channel at any time can be fully determined. Therefore, the fouling development in a full-scale RO process can be well modeled. A sample calculation is given below:

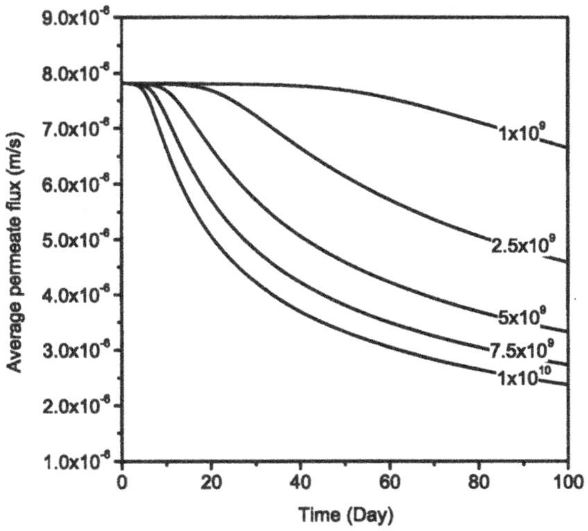

Figure 12.1.6: Change of average permeate flux in a 6-m membrane channel for different k-values

Membrane fouling characterization

Fouling will inevitably result in increase in membrane resistance under any time and therefore change in membrane resistance can be a fault-free indicator for membrane fouling. That means if the membrane resistance keeps constant over operating period, it is safe to say there is no fouling. However, because membrane resistance is a distributive parameter that has different value at different location due to uneven membrane fouling, the direct use of the localized membrane resistance to indicate membrane fouling is not convenient and it is unpractical to find out membrane resistance at every point in the membrane channel. To overcome this difficulty, a lumped (collective) parameter, *filtration coefficient*, was defined

$$F = \frac{1}{u_0 H} \int_o^L \frac{dx}{R_m(x)} \tag{11}$$

where F is the filtration coefficient, u_0 is the feed flow velocity, H is the channel height, and $R_m(x)$ is the localized membrane resistance. For a given RO process, F is a function of membrane resistance solely. Any change of membrane resistance at any point of the membrane channel will be reflected in F.

With the filtration coefficient, the recovery can be linked to the net driving pressure with the following equation:

$$R = F(\Delta p - \Delta \pi) \tag{12}$$

Eq (12) showed that the filtration coefficient would be the slope when the recovery is plotted versus the driving pressure. Because the information of the localized membrane resistance is unavailable, the filtration coefficient usually cannot be calculated directly from Eq. (11). Instead, the filtration coefficient can be easily determined from Eq. (12) with a simple pilot test, by finding the slope in the R vs. (Δp–$\Delta\pi$) plot (Fig. 12.1.7). (The testing has to be done in a low pressure such that the full-scale RO is operated in the mass transfer controlled regime.)

From Eq. (9), it can be seen that the filtration coefficient F decreases as the overall membrane resistance increases. Therefore, membrane fouling in a full-scale RO process can be described by a series of straight lines with different slopes at different times (Fig. 12.1.8). For a given driving pressure (the vertical line in the figure), recovery of the full-scale RO process (the intersection point of the recovery line with the vertical pressure line) decreases with time as membrane fouling is going on.

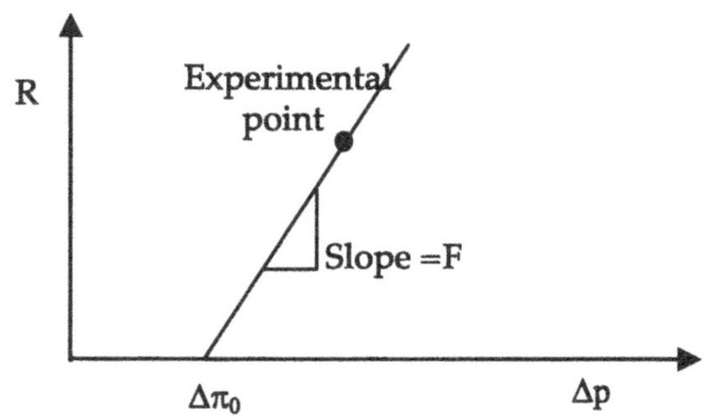

Figure 12.1.7: Filtration coefficient is the slope of the line of R vs Δp

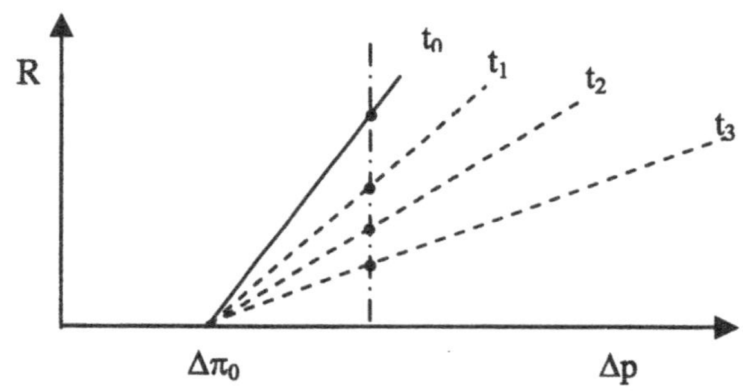

Figure 12.1.8: Fouling can be indicated by the straight lines with different slopes

To our best knowledge, it is the first time that this complex but rather important fouling phenomenon in the full-scale RO processes is explained with the shifting between controlling mechanisms of the process. It can provide profound insights rather than phenomenological descriptions of the processes. For example, the use of filtration coefficient for the evaluation of cleaning efficiency can provide more accurate information than the use of average permeate flux.

Concluding Remarks

One conventional practice in both membrane module and process designs is to facilitate the use of large membrane area. The spiral-wound RO modules squashes as large as possible membrane area into a limited space. In the full-scale RO processes, long pressure vessels holding 6~7 modules are standard design and 2 or 3 stages of the long pressure vessels are connected in series. These measures were necessary for the old generations of RO membranes that had permeability orders small in magnitude than the membranes we are using currently. It can be proven that even with these measures to increase membrane areas for a given feed water supply, the processes with old generations of membranes are still controlled by mass transfer. This result justifies the design strategy as described above for the old generations of membranes.

With the emergence of the thermodynamic restriction with the new generation of highly permeable RO membranes, a new strategy of module and process design is required.

The emergence of thermodynamic restriction itself reveals there is more than needed membrane area in the processes. It is an indicator that the capacity of RO membranes has not been fully utilized because there is not sufficient feed water supply to the entire membrane. This is evidenced by the hydraulic imbalanced problem as reported in the literature. It was reported with the highly permeable RO membranes, the last a few modules do not have water to produce because the net driving force has been significantly reduced. This problem can be improved by increase the channel height and reduce the total membrane area in the modules.

It seems that the module and process design is a bottleneck for further improvement of membrane permeability. If the design cannot even utilize the current capacity of the RO membranes, there is no need to further improve the membrane permeability. A new design strategy must be developed to fully use up the capacity of the current RO membranes and leave some room for further improvement.

References

J.S. Taylor, E.P. Jacobs, Reverse osmosis and nanofiltration, in Water Treatment: Membrane Processes, American Water Work Association Research Foundation, McGraw Hill, New York, 1996. 2.

M. Mulder, Basic Principles of Membrane Technology 2rd ed., Kluwer Academic Publishers, London, 1996. 3.

M. Taniguchi and S. Kimura, Estimation of transport parameters of RO membranes for seawater desalination, AIChE Journal, 46 (10) (2000) 1967-1973. 4.

M. Taniguchi, M. Kurihara, and S. Kimura, Behavior of reverse osmosis plant adopting a brine conversion two-stage process and its computer simulation, Journal of Membrane Science, 183 (2001) 249-257. 5.

N. M. Al-Bastaki and A. Abbas, Modeling an industrial reverse osmois unit, Desalination, 126 (1999) 33-39. 6.

N. M. Al-Bastaki and A. Abbas, Prediction the performance of RO membranes, Desalination, 132 (2000) 181-187. 7.

L. Song, S. K. Hong, J. Y. Hu, S. L. Ong, and W. J. Ng, Simulations of full-scale reverse osmosis membrane process, Journal of Environmental Engineering, ASCE, 128 (10) (2002) 960-966. 8.

M. Wilf, Design consequences of recent improvements in membrane performance, Desalination, 113, (1997), 157-163. 9. J.E. Nemeth, Innovative system designs to optimize performance of ultra-low pressure reverse osmosis membranes, Desalination, 118 (1998) 63-71. 10.

L. Song, J. Y. Hu, S. L. Ong, W. J. Ng, M. Elimelech, and M. Wilf, Performance limitation of the full-scale reverse osmosis process, Journal of Membrane Science, 214 (2003) 239–244

12.2 Innovative Simulation for Membrane Bioreactor and the Quantitative Analysis of Membrane Performance

Hiroo Takabatake[1], Masatsugu Uehara[1], Yohito Ito[1], Kanao Otake[1], Masahiro Henmi[1] and Tadahiro Uemura[1]

Toray Industries, Inc. Shiga Japan[1]

Abstract

Membrane Bioreactor (MBR) has been focused as the strong candidate of decentralized wastewater reclamation system because it has very many advantages. In order to improve the MBR operation, we have constructed the simulation method to predict the operation results based on the two important key parameters. One is a detachment parameter and the other is a resistance parameter. The simulation results of MBR operation were very close to the actual operational results, indicating that the simulation results were considered to be highly reliable. Then, the simulation was applied to the evaluation of the several kinds of membrane performance, and the relationships between membrane performance and morphology on the membrane surface were investigated. The results show that the increase in the roughness of membrane surface causes the detachment parameter to be decreased and that the increase in membrane pore size causes the resistance parameter to be increased. This result indicated that the smoother membrane surface and the smaller pore size lead the membrane to the better performance for sludge filtration.

Introduction

Recently, the establishment of decentralized wastewater reclamation systems is expected as promising systems from the aspect of water circulation. Membrane Bioreactor (MBR), of which main instrumental feature is the solid/liquid separation of activated sludge by the membrane, has been focused as the strong candidate of them because it has very many advantages. For example, it has smaller footprint, higher effluent quality and lower amount of excess sludge production than the process with the solid/liquid separation by gravitational sedimentation.

One of the most important points for stable operation of MBR is the regulation of fouling, which is the phenomenon declining the performance of membrane filterability. It is considered that the fouling analysis is the important point not only for the stable operation of MBR but also for the improvement of MBR technology. For them, several simulation models have been reported to predict the MBR operation results[1-6]). However, most of them were just prediction or reproduction of MBR operational results, and few examples applied to the MBR operation or contributed to technology improvement of MBR.

We have constructed the simulation method, which is available not only to predict the change of MBR operational results but also to evaluate the membrane performance in sludge filtration.

Nomenclature

A	effective membrane area [m ^{2}]	Zave	average height of Z axis in scanning area [ì m]
J	flux [m/s]	Zn	height of Z axis [ì m]
P	concentration of P component [gC/m ^{3}]		
Pm	amount of P attached on the membrane surface [gC/m ^{2}]	áp	resistance parameter of P [m/gC]
ÄP	pressure [Pa]	áx	resistance parameter of X [m/gC]
R	filtration resistance [1/m]	ãp	detachment parameter of P [1/m/s]
Rm	initial filtration resistance [1/m]	ãx	detachment parameter of X [1/m/s]
RMS	roughness of membrane surface [ì m]	çp	inverse of the dencity of P [m ^{3}/gC]
S	concentration of S component [gC/m ^{3}]	çx	inverse of the dencity of X [m ^{3}/gC]
t	time [s]	ëp	friction parameter of P [1/Pa]
V	volume of activated sludge [m ^{3}]	ëx	friction parameter of X [1/Pa]
Vo	initial volume of activated sludge [m ^{3}]	ô	washing stress in the membrane surface [-]
X	concentration of X component [gC/m ^{3}]	ì	viscosity of the permeate [Pa s]
Xm	amount of X attached on the membrane surface [gC/m ^{2}]		

In this paper, the reliability of this simulation results were checked and the relationship of membrane performance and the morphology of membrane surface were studied. Through this study, it is discussed what feature of membrane surface lead a membrane to a good performance of sludge filtration.

Materials and Methods

Simulation development

The simulation method was developed to simulate the change of membrane flux, trans-membrane pressure and filtration resistance during the sludge filtration. The fundamental model equations and the simulation method are shown as follows.

Definition of activated sludge

It was defined that activated sludge consists of solid component (X), particle component (P) and soluble component (S): "X" is the sediment by a centrifugation, "P" is neither the sediment by a centrifugation nor the permeate through membrane, and "S" is the permeate through membrane. The concentration of X component was calculated by an equation X [mgC/l] = MLSS [mg/l] * 60/113, where MLSS is the measurement value of MLSS [mg/l], and 60/113 is the conversion ratio from a unit of [mg/l] to [mgC/l]. The concentration of S component was determined as TOC concentration of membrane permeate water, and the concentration of P component was to be the difference of the TOC concentration of the centrifuge supernatant and the membrane permeate.

Equations of filtration resistance

The amount of X and P component attached on the membrane surface was calculated according to the following equations.

$$dXm/dt = X\,J - \gamma x\,(\tau - \lambda x\;\;P)\,(\eta x\,Xm + \eta p\,Pm)\,Xm \quad (\tau - \lambda x\;\;P >= 0)$$
$$dPm/dt = P\,J - \gamma p\,(\tau - \lambda p\;\;P)\,(\eta x\,Xm + \eta p\,Pm)\,Pm \quad (\tau - \lambda p\;\;P >= 0)$$

, where X and P are the concentration of X and P component respectively [gC/m^3], Xm and Pm are the amount of X and P component attached on the membrane surface respectively [gC/m^2], t is time [s], J is flux [m/s], γx and γp are the detachment parameter of X and P respectively [1/m/s], τ is the washing stress on the membrane surface [-], λx and λp are the friction parameter of X and P [1/Pa], ΔP is the trans membrane pressure [Pa], ηx and ηp are inverse of the density of X and P component respectively [m^3/gC]. The above equations represent the balance of attaching rate onto membrane surface and detaching rate from membrane surface, caused by the washing stress generated by the cross flow of water and bubble.

The filtration resistance was calculated by the following equations.

$$R = Rm + \alpha x\,Xm + \alpha p\,Pm$$

, where R is the filtration resistance [1/m], Rm is the initial resistance [1/m], αx and αp are the resistance parameter of X and P respectively [m/gC]. And R has the following relationship;

$$J = \Delta P \;/\; (\mu\,R)$$

, where μ is viscosity of the permeate [Pa s].

Material balance on X or P supplies following equations

$$XV + AXm = \text{constant}$$

$$PV + APm = \text{constant}$$

,where V is volume of activated sludge [m^3] and A is effective membrane area [m^2]. In the case of dead-end filtration, following equation is necessary.

$$V = Vo - \text{integral}\,(A\,J)\,dt$$

,where Vo is initial volume of activated sludge [m^3].

Calculation method

In order to solve above equations, MATRAB (Mathworks) was applied as a calculation soft, using Runge-Kutta method for integration.

Evaluation of membrane performance by the simulation

The performances of 4 types of flat sheet membrane listed in Table 12.2.1 were evaluated by dead-end filtration tests followed by their simulations. The apparatus equipment of dead-end filtration test was shown in Fig. 12.2.1. For this filtration test, each of membranes listed in Table 12.2.1 was cut to the circle (effective membrane area was 4.1cm^2) and set in the membrane holder of stirring cell after conditioned by the submergence in 20% ethanol solution for more than 2 hours. Before the filtration test of activated sludge or its supernatant, pure water filtration was done for acquiring the data of membrane resistance. The filtration test of activated sludge or its supernatant was carried out under 20kPa of filtration pressure with the occasional relaxation and with agitation by the magnet rotation at 600rpm. During filtration, the permeate amount data were automatically taken into PC during filtration every 1 second. From these data, the changing of filtration resistance was calculated, and all parameters dominating the characteristics of membrane or activated sludge were determined from the changing of filtration resistance.

Table 12.2.1: Flat sheet membrane applied to the evaluation of filtration performance

Membrane	A	B	C	D
Nominal pore size	0.08μm	0.4μm	0.1μm	0.1μm
Water contact angle	64°	47°	43°	88°
Surface structure	smooth	network	network	network
Membrane supplier	Toray	X	Y	Y

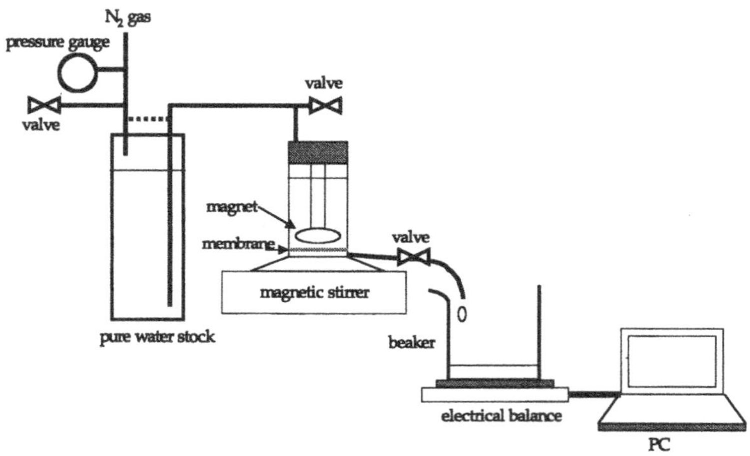

Figure 12.2.1: The apparatus equipment of dead-end filtration test

Analysis of membrane morphology

The morphology of membrane surface was analyzed using AFM (Atomic Force Microscope) equipment (Digital Instruments, Nanoscope IIIa) with contact scanning mode. The roughness of membrane surface was calculated as the following equation.

$$RMS = (\Sigma ((Z_n - Z_{ave})^2 / N))^{1/2}$$

,where RMS is roughness of membrane surface [μm], Z_n is height in Z axis [μm] and Z_{ave} is average height in scanning area [μm].

Results and Discussion

Reliability check of the simulation results

The MBR filtration test was carried out in MBR (effective volume: $2.4m^3$) treating industrial wastewater with PVDF flat-sheet membrane (the same as membrane A in Table 12.2.1). In this MBR filtration test, membrane permeate was obtained by water head difference, and the elevation water head of permeate exit was changed at fixed interval and permeate flow rate was monitored and recorded during filtration. The flow rate observed in the MBR filtration test was simulated after the determination of all parameters dominating the characteristics of membrane and sludge from the result analysis of dead-end filtration test. In this simulation, only τ, dependent with reactor construction and bubbling condition for scouring membrane surface, was changed as minimizing the difference between the simulation value and the measurement value. The simulation results are shown in Fig. 12.2.2 together with the measurement results. The simulation results were very close to the measurement results except for the sharp flushing in the start of filtration, indicating that the simulation results are highly reliable.

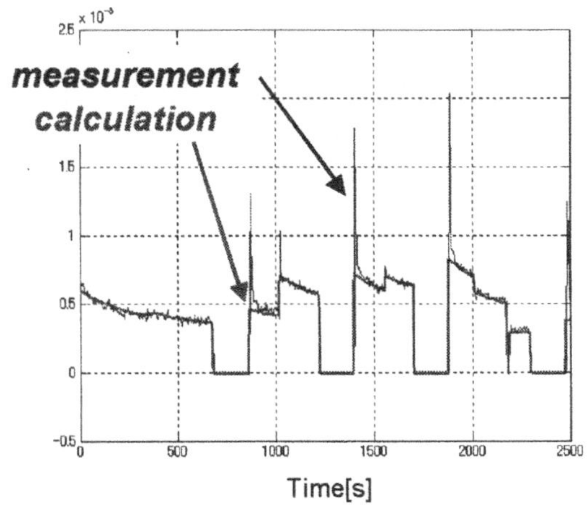

Figure 12.2.2: The flow rate comparison between measurement value and calculation value.

Evaluation of membrane performance by simulation

The 4 types of membrane listed in Table 12.2.1 were evaluated its sludge filtration performance by the simulation. These membranes are variegated in pore size, water contact angle and in surface structures, smooth or network (examples of SEM (Scanning electron microscope) photo of them were shown in Fig. 12.2.3).

By using membranes, dead-end filtration tests were carried out with the same activated sludge. The example of the test result (membrane C for sludge filtration) was shown in Fig. 12.2.4 together with the simulation results. In this test, the sludge filtration was temporary stopped for 1 minute when 0.012 [m] of permeate volume per membrane area was obtained, observed sudden decrease in filtration resistance. The same test was done used centrifuge supernatant. Firstly, the parameters relating with P component, which are αp, γp and λp, were determined as the simulation result got close to the measurement result of test in the supernatant. After that, the parameters relating with X component, which are αx, γx and λx, were determined as the simulation result (αp, γp and λp determined above were utilized to this simulation) got close to the measurement result of test in the sludge. Figure 12.2.4 indicates that the simulation result was very close to the measurement result, supporting that the determined values were reliable.

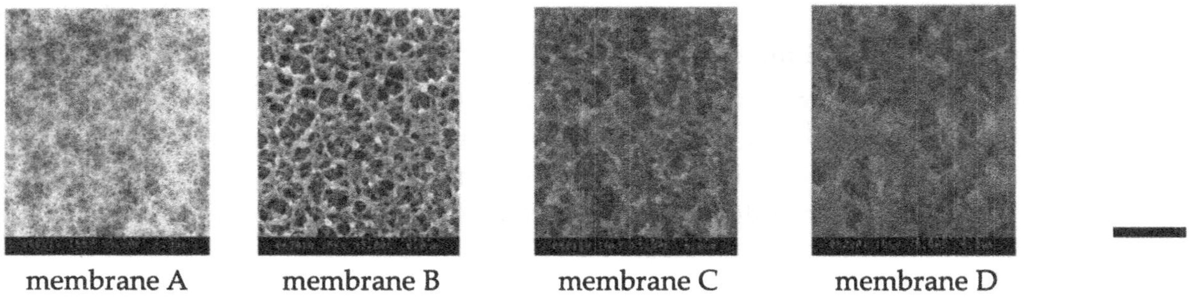

Figure 12.2.3: Examples of SEM photo of membrane with "smooth" surface structure (membrane A) and "network" surface structure (membrane B, C and D). The black bar length indicates 6.0μm.

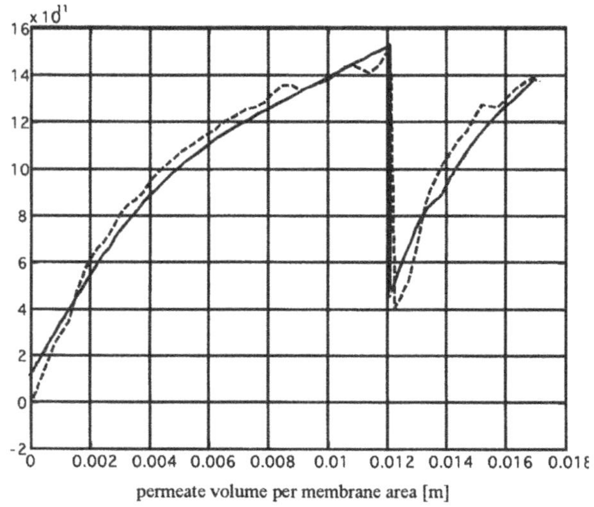

Figure 12.2.4: The example of the result of dead-end filtration test (membrane C for sludge filtration) together with the simulation result. The continuous line is the simulation result and the dashed line was the measurement results.

Among the parameters determined above, the resistance parameter ·x and the detachment parameter Áx are much important to discuss the membrane performance because they are the most sensitive to the simulation results. The resistance parameter represents the resistance generated by a unit amount of sludge component attached on the membrane surface, and the detachment parameter represents the easiness of sludge component detaching from the membrane surface. Therefore, the higher detachment parameter means the more difficult for sludge attaching on a membrane surface, and the lower resistance parameter means the lower resistance generation from the sludge attached on a membrane surface.

The detachment parameter and resistance parameter determined on each membrane are shown in Fig. 12.2.5. Membrane A is the lowest in resistance parameter and the highest in detachment parameter. Especially, the detachment parameter of membrane A is significantly higher than others. It is indicated that the main feature of membrane A is considered as the difficulty for sludge component to be accumulated on the membrane surface, making membrane A difficult to be fouling.

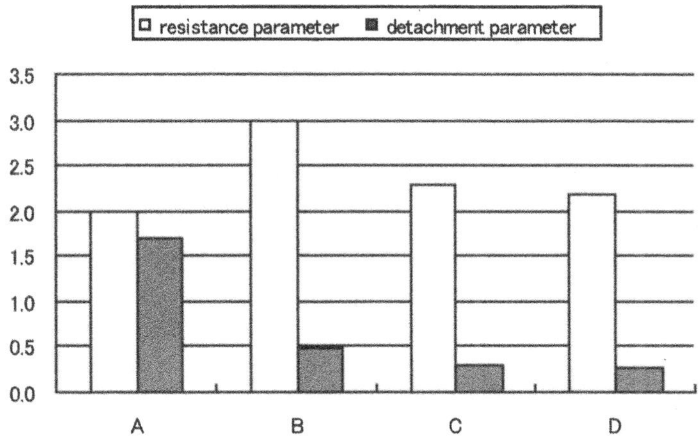

Figure 12.2.5: The detachment parameter and resistance parameter determined on each membrane.

Relationships between membrane performance and morphology of membrane surface

Relationships between membrane performances, represented as resistance parameter and detachment parameter, and morphology of membrane surface, focused on the roughness of membrane surface and nominal pore size. The detachment parameters and the resistance parameters of each of membranes are listed in Table 12.2.2 together with the roughness of membrane surface (RMS) and the nominal pore size of each membrane. The relationship between detachment parameter and the roughness of membrane surface and between resistance parameter and the pore size of membrane were shown in Fig. 12.2.6. The increase in RMS tends to decrease the detachment parameter, and the increase in pore size tends to increase the resistance parameter. Especially the relationship between RMS and the detachment parameter is much significant. These tendency means that the increase in the RMS causes the detachment parameter to be decreased and that the increase in membrane pore size causes the resistance parameter to be increased, indicating that the smoother membrane surface and the smaller pore size lead the membrane to the better performance for sludge filtration.

Table 12.2.2: The morphology of membrane surface (nominal pore size and roughness on membrane surface (RMS)) and membrane performance (detachment parameter and resistance parameter).

Membrane	A	B	C	D
RMS [μm]	0.062	0.180	0.279	0.417
Nominal pore size [μm]	0.08	0.4	0.1	0.1
Detachment parameter (γ) [10^3/m/s]	1.70	0.50	0.30	0.28
Resistance parameter (α) [10^{12}m/gC]	2.0	3.0	2.3	2.2

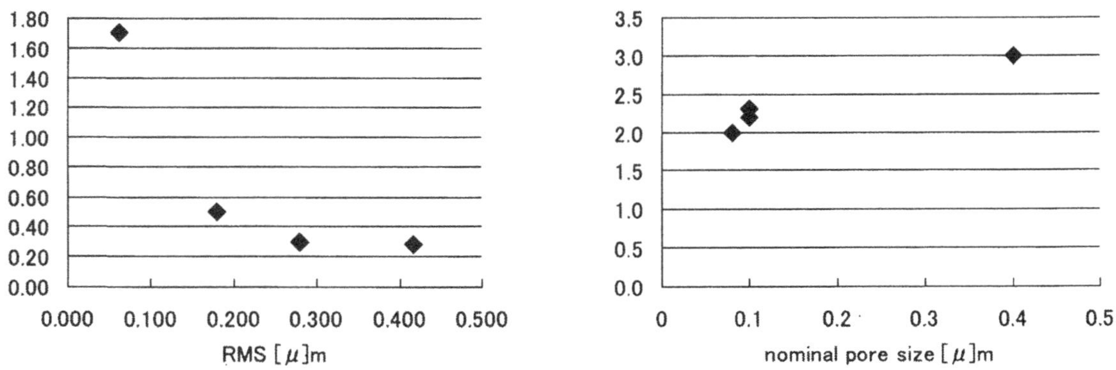

Figure 12.2.6: The relationship between detachment parameter and the roughness of membrane surface (left) and between resistance parameter and the nominal pore size of membrane (right).

Conclusions

We have constructed the simulation method to predict the result of MBR operation. The simulation results of MBR filtration test were very close to the measurement results of it, indicating that the simulation result is reliable. This simulation method is available to evaluate the membrane performance of sludge filtration. In this evaluation, the detachment parameter and resistance parameter are much important. The performance evaluation of 4 types of membrane represented the feature of these membranes, and the discussion in combination with the analysis of the morphology of membrane surface make it cleared the tendency that the increase in the roughness of membrane surface decreased the detachment parameter and the increase in the nominal pore size decrease the resistance parameter. It means that the smoother membrane surface and the smaller pore size lead the membrane to the better performance for sludge filtration.

References

Nagaoka H., Yamanishi S., Miya A. (1998): Modeling of Bio-Fouling by Extracellular Polymers in a Membrane Separation Activated Sludge System, *Water Science and Technology*, **38**(4-5), 497-504.

Lim A.L. and Bai Renbi (2003): Membrane Fouling and Cleaning in Microfiltration of Activated Sludge Wastewater, *Journal of Membrane Science*, **216**, 279-290.

Lu S.G., Imai T., Ukita M., Sekine M., Higuchi T., Fukagawa M. (2001): A Model for Membrane Bioreactor Process Based on the Concept of Formation and Degradation of Soluble Microbial Products, *Water Research*, **35**(8), 2038-2048.

Lee Y., Cho J., Seo Y., Lee W.J., Ahn K.H. (2002): Modeling of Submerged Membrane Bioreactor Process for Wastewater Treatment, *Desalination*, **146**, 451-457.

Wintgens T., Rosen J., Melin T., Brepols C., Drensla K., Engelhardt N. (2003): Modeling of a Membrane Bioreactor System for Municipal Wastewater Treatment, *Journal of Membrane Science*, **216**, 55-65.

Cho J., Song K.G., Ahn K.H. (2005): The Activated Sludge and Microbial Substances Influences on Membrane Fouling in Submerged Membrane Bio-reactor: Unstirred Batch Cell Test, *Desalination*, **183**, 941-945.

12.3 Oxidation of Hydrated Organics in Paper Mill Wastewater

Katherine S. Taylor[1], and Jeffrey Sturman[1], and Charles Bucknell[2]

[1] Murdoch University Environmental Technology Centre
[2] Amcor Fibre Packaging Mill Spearwood

Abstract

Recyled paper mill wastewater is currently treated and irrigated on site at Amcor Fibre Packaging Mill Spearwood, in Western Australia . The residual organics in wastewater create odour and filter fouling problems. In this study, parameter spaces of the following potential treatments were investigated: enzymes (cellulase mixtures), microbial digestion, and wet oxidation (150°C and 540- 560kPa with combinations of NaOH, H_2O_2, and CH_3COOH). The success of treatments was assessed by comparing the chemical oxygen demand (COD) and filterability before and after treatment.The enzyme treated wastewater samples showed no significant difference to the control, possibly due the presence of in situ bacterial populations or cooked starch. Aerobic bacterial digestion reduced filterability and COD reduction (maximum 17%) was unsatisfactory. Wet oxidation significantly increased the filterability and COD reduction of wastewater. Cooking time was a significant factor. Up to 79% reduction in COD was acheived after 30 minutes of wet oxidation. Acetic acid and 10 minutes of wet oxidation removed 83% of COD, but also corroded the test apparatus. In this pilot study, wet oxidation was the most effective wastewater treatment tested , but further investigation into the process parameters are needed.

Introduction

The Amcor Fibre Packaging Spearwood Mill ("Amcor"), in Western Australia, produces a range of packaging materials. The mill uses 10 to 15 kilolitres of water per tonne of paper produced from recycled paper. The cellulose is removed by clarification and filtration (~ 95% efficiency) prior to beneficial reuse by onsite irrigation. The final effluent contains hydrated cellulose and has a temperature of 40°- 50° C. This wastewater is distinct from whitewater, which is the water which drains directly from the papermaking process, although both can contain cellulose fibres. The residual organics in wastewater have fouled filters and have clogged the irrigation system. Soil pores can also been clog, leading to soil saturation, anaerobic conditions, and odour.

We hypothesised that the residual organics in the wastewater could be reduced by oxidation using biological or chemical treatments. Therefore, the project objective was to identify and test oxidising agents to remove residual organics from the wastewater.

Literature Review

Cellulose fibres in wastewater

Cellulose fibres have high tensile strength and inherent hydrogen bonding ability. They are water insoluble, hydrophilic, variable in size, chemically stable, and are relatively colourless. The hydrophilic nature of cellulose means that the fibres absorb water and swell, becoming dispersed throughout the solution (Smook, 1988). When recycled, the fibres loose many external fibrils, which reduces their cohesiveness. The structure of the fibre is more tightly bound, so the surface is more difficult to fibrillate. The fibres become shorter, and stiffer, which makes them prone to further shortening (Clark, 1985, Smook, 1988). These changes make the fibre less likely to be retained with other fibres in the papermaking process and consequently they become part of the wastewater.

Outline of some possible treatment processes

1. Enzyme treatment. Compared to other biological treatments, the advantages of using enzymes to treat wastewater are that they can: operate at a range of contaminant concentrations, pH, temperature and salinity; target specific contaminants and the process can be easily controlled. In the paper industry, the cost of enzymes has prohibited the industrial production of sugars from cellulose in the past. However, they can theoretically be used to treat wastewater. It been demonstrated that residual paper inks in whitewater do not inhibit enzyme action (Karam & Nicell, 1997).

2. Microbial digestion. One aerobic process that has been used to treat paper mill whitewater is the Kaldness suspended polystyrene carrier process, which is similar to activated sludge (Dalentoft & Thulin, 1997). Another process is the membrane bioreactor, which Tardiff and Hall (1997) report to have removed 84% of COD from whitewater at temperatures of 40-55°C and at a hydraulic retention time of 2.8 days. A combination of anaerobic and aerobic treatment is used in a "zero discharge" paper recycling mill in the Netherlands (Habets & Knelissen, 1997). Anaerobic digestion is not common in the paper industry due to whitewater's high sulphur content, which forms H_2S gas (Smook, 1988).

3. Wet air oxidation. Wet air oxidation produces inorganic salts or simpler forms of the organic molecules which were originally present. Forms of wet air oxidation have been used to successfully treat paper mill whitewater. Minowa & Inoue (1998) converted cellulose into methane gas at temperatures of 350°C and pressure of 18mPa. Clark (1985) hydrolysed cellulose with dilute acid at temperatures above 150°C, as a pre- treatment before fermenting it to alcohol. Verenich *et al.* (2001) removed COD from paper mill thermomechanical pulp (waste from refining woodchips) by coagulation with $Fe_2\ SO_4$, then wet oxidation. The iron acted as a catalyst and improved the wet oxidation.

Amcor's wastewater

The wastewater is opaque and has a brown- grey colour. Aerobic wastewater has an approximately neutral pH. It contains cellulose, starch, bacteria, and salts. In addition, it possibly contains the following recycled pulp contaminants: adhesives, polystyrene, plastic chips, plastic films, resins, latex, waxes, asphalt, and non-cellulose fibres (Smook, 1988). The wastewater may vary in composition depending on paper products being manufactured and the quality of the recycled paper entering the system. Results from preliminary investigations of this study indicate that after wastewater treatment, further settling of organics can be achieved. This observation led to the following conclusions: firstly, that not all organics in solution were fully hydrated; and secondly, the amount of time a sample has settled for before filtration must be known in order to achieve consistent results.

Materials and Methods

The success of all wastewater treatment methods tested was evaluated by changes to wastewater COD and filterability. COD was tested by Hach method 8000: reactor digestion and then analysis with a spectrophotometer. Filterability was tested by creating a vacuum in a 1L graduated cylinder with a vacuum pump. The sample was filtered through 2 layers of filter paper (90mm diameter Whatman GF/C supported by Whatman 541). The volume filtered was recorded at the time intervals: 30 seconds, 1 minute, 2 minutes, 3 minutes, 4 minutes, and 5 minutes. The rate of filtration is referred to here as "filterability". Samples that were "mixed" were shaken and then filtered within 5 minutes of mixing. Samples that were "settled" were settled for 1 hour (unless otherwise specified) and were then filtered. Samples used in different tests came from different wastewater batches. However, the samples used in any one test came from the same batch.

Enzyme testing

Five commercially available enzyme samples were used. They were Cellulase ESP 136 (Cellulase-Xylanase, Glucanase), Validase TRL (Cellulase), Cellulase DP 132 (Cellulase), Multifect A40 (Cellulase, Hemicellulase, Xylanase, Mannanase) and Enzymatic. Note that although the author misunderstood "Enzymatic" to be an enzyme mix, *it is actually a nutrient supplement for microbial cultures*. The enzymes were added to 1L samples of wastewater, which was then left in a heat bath at 40-50°C or at ambient temperature (20-30 °C). The samples were not stirred or aerated. Several different enzyme doses were tried (20μL, 1mL, 10mL). The effect of acidifying with acetic acid was tested.

Microbial treatment

3L of wastewater was placed in 5L buckets. The buckets were aerated with an aquarium pump at 2L air per minute. The bacterial culture ("BC") included aerobic and facultative anaerobic bacteria selected for the paper –pulp industry (Motherwell, 2006) and was added at concentrations of 1:1000 and 1:10 000.

Wet oxidation

1L samples of wastewater were "cooked" in a custom made reaction vessel (the "reactor") consisting of a 1.5L steel cylindrical vessel with steam and air inlets near the base, an inlet for introducing the sample, a steam/ air outlet valve at the top, and a sample outlet tap at the lowest point. The air flow rate was approximately 3L per minute. A perforated copper pipe was installed inside the reactor in line with the air and steam inlets to aid the mixing of air through the sample. The cooking temperature was about 150°C (+- 4°C) and the pressure ranged from 540- 560 kPa (+- 20 kPa). The cooking times tested ranged from 5 to 30 minutes. In addition to the cook time, samples spent 4-6 minutes in the reactor reaching operating conditions and 4-6 minutes cooling. The effects of adding the following reagents were tested: hydrogen peroxide (H_2O_2), sodium hydroxide (NaOH), and acetic acid (CH_3COOH).

Results

All results which are described here as "significant" were statistically significant at the 95% level. The difference in means of two samples was determined by t-tests. The difference in means of several samples was determined by one-way ANOVA. The statsistical package used was SPSS 12.0.

Enzymes

A dataset was complied from 5 different enzyme tests. The mean filterability of samples treated with "Enzymatic" was less than the mean filterability of samples treated with the other enzymes and also the control. There was no significant difference between the other enzymes and the control

The filterability of acidified samples treated with enzymes in the heat bath for one day was not significantly different to the control. After 4 days, the filterability was still not significantly different to the control and neither was the COD reduction. The COD reduction of samples in the heat bath was not significantly different to the untreated acidified sample kept in the fridge.

In another test, the COD reduction and filterability of samples treated on the bench was not significantly different to the filterability of samples treated in the heat bath.

Table 12.3.1: Reduction in COD after filtration and enzyme treatments at 40-50 °C (heat bath) compared to 20-30°C (ambient temperature). Reduction in COD after treatment with acetic acid and enzymes.

Treatment	COD reduction (%) after 1 day in heat bath	COD reduction (%) after 1 day in ambient	Acidified* samples COD reduction (%) after 4 days heat bath
Original (fridge storage)			44.0
Control	-	15%	46.6
Validase TRL	-	-	45.0
Enzymatic	2%	3%	47.6
Multifect A40	6%	3%	45.4
Cellulase ESP 136	10%	15%	46.0
Cellulase dp 132	7%	12%	42.4
Mean	**5%**	**8%**	**45.3**

**Filtrate COD reduction was adjusted for COD of acetic acid*

Microbial digestion

The highest filterability increase was gained by anaerobic digestion for 5 days. However, the treatment increased the COD of the samples. The highest COD reduction was achieved by aerobic digestion for 5 days, but treatment also decreased the filterability. After 10 days of aerobic digestion, the filterability of samples was very low, and the suspended solids were resistant to settling and centrifuging.

Generally, the filterability of mixed samples was significantly lower at longer retention times. There was no significant difference in filterability between the samples with bacterial culture added and those without.

Table 12.3.2: Results of microbial digestion tests with and without added bacterial culture (BC)

The settling time for filtration was 1 hour. COD of untreated wastewater was compared to treated filtrate.

Date	1= air 0=no air	BC (mL/L)	Retention time (days)	Filtration mixed (mL)	Filtration settled (mL)	Nutrient	COD (%) reduction
03 Feb	1	0	3	145	.	.-	
	1	1	3	140	.	.-	
06 Feb	1	0	1	425	.	.-	
	1	1	1	447	.	.-	
untreated	0	0	0	375	.	.-	
10 Feb	1	0	10	.	25	.-	
	1	1	10	.	150	.-	
14 Feb	1	0	1	240	270	.-	12
	1	1	1	270	280	.-	11
	0	0	1	180	270	.-	10
	0	1	1	200	220	.	11
21 Feb	1	1	1	210	220	+	
	1	1	1	240	250	-	
untreated	0	0	0	210	.	.-	
22. Feb	0	0	5	.	460	.-	-16
	0	0.05	5	.	800	.-	-15
	1	0	5	.	130	.-	16
	1	0.05	5	.	90	.-	17

Wet oxidation

After the wastewater was cooked, it was clearer and darker. Precipitates were formed in the samples cooked with NaOH (grey) and acetic acid (predominately red). The most important factor in predicting increases in filterability and COD reduction was cooking time. Figure 12.3.1 below shows this relationship.

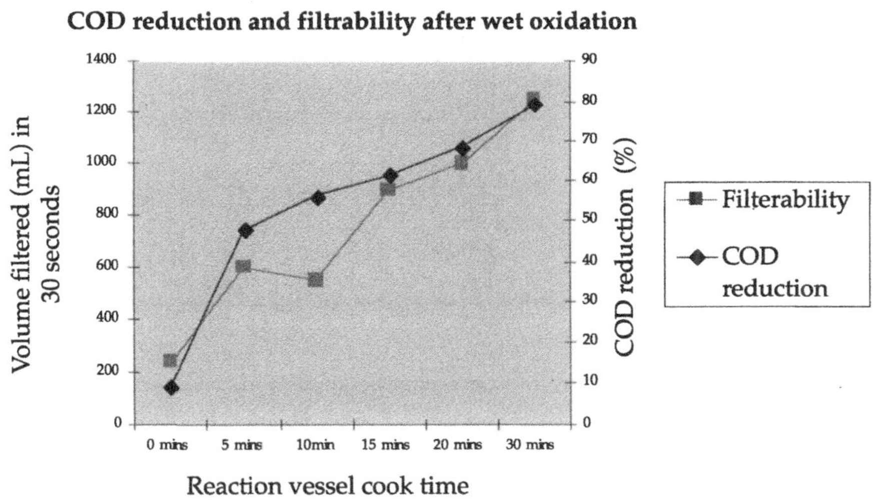

Figure 12.3.1: Volume of sample filtered in 30 seconds and COD reduction for samples cooked in the reactor without reagents February 27th.

The results from the test run (February 24th) were excluded from statistical analysis. When cook time and reagents were used as variables in a linear regression model of COD reduction, the significant variables were cook time, acetic acid and hydrogen peroxide, which all increased COD reduction. NaOH was not significant. The effect of H_2O_2 was not significant when it was used alone. The reduction of COD caused by filtering untreated wastewater had a mean of 13%, and a range of 9%- 18%.

When filterability was tested using a similar linear model, cook time and H_2O_2 were once again statistically significant. The presence of CH_3COOH and NaOH significantly decreased filterability. Again, without other reagents, there was no significant effect from H_2O_2.

Table 12.3.3: Wet oxidation treatments of wastewater, the reduction in COD, and volume of treated wastewater which could be filtered in 30 seconds. The COD of each batch is in the left column.

Date	Time in reactor (150°C, 540-560kPa) and reagents added	Volume filtered (mL)	COD of filtrate	Reduction COD (%)
24-Feb	10 minutes	460		
	20 minutes	650		
	30 minutes	540		
	30 minutes + NaOH	270		
	30 minutes + H_20_2,NaOH	850		
	30 minutes + H_20_2	680		
	No treatment (0 minutes)	130		
27-Feb	No treatment (0 mins) COD= 2530	240	2290	9.49
	5 minutes	600	1320	47.8
	10 minutes	550	1110	56.1
	15 minutes	900	980	61.3
	20 minutes	1000	800	68.4
	30 minutes	1250	530	79.1
28-Feb	No treatment (0 mins) COD= 3190	220	2600	18.5
	5 minutes	1000	1442	54.8
	5 minutes +H_20_2	900	1593	50.1
	5 minutes + H_20_2+2mLNaOH	750	1420	55.5
	5 minutes + H_20_2+ 8mLNaOH	340	1528	52.1
	5 minutes + 2mLNaOH	420	1541	51.7
	10 minutes	1000	1160	63.6
	10 minutes + H_20_2	920	1280	59.9
	10 minutes + H_20_2+2mL NaOH	750	1227	61.5
	10 minutes + H_20_2+ 8mL NaOH	290	1244	61
	10 minutes + 2mL NaOH	900	1260	60.5
3-March *	5 minutes	850	1320	45.2
	10 minutes	860	1250	48.1
	5 minutes + H_20_2 10mL, acetic 2mL	930	3000 -2163 =837	65.3
	10 minutes + H_20_2 10mL, acetic 2mL	670	2160 -2163 =-3	100
	5 minutes +acetic 2mL	270	3290 -2163 =1127	53.2
	10 minutes +acetic 2mL	560	2580-2163 =417	82.7
	No treatment (0 mins) COD=2410	150	2120	12

**COD of 2mL acetic acid=2163*

Discussion

There was difficulty in estimating the destruction of cellulose due to the other organic compounds in the wastewater that contribute to COD. A cellulose specific test would have helped the interpretation of results. Similarly, the tests used here do not reveal which species of organic compounds were removed by filtration. An increase in filterability would not necessarily correspond to a decrease in cellulose fibres.

However, some non-cellulose organics in the wastewater cause the same problems as cellulose i.e., filter fouling and odour. Therefore, although testing COD and filterability gave non-specific results, they were still meaningful tests.

Enzyme catalysts

The difference in COD and filterability between the control and enzyme treated samples were not significant, regardless of enzyme concentration, retention time, acidification, or retention temperature. The exception was "Enzymatic", which gave significantly lower filterability. The difference can be attributed to increased microbial growth in the presence of a nutrient source (Enzymatic). None of the samples were sterilised, so microbial digestion occurred.

When the enzyme supplier was consulted, he suggested that the enzyme action was inhibited by the presence of other organics in the wastewater, namely cooked starch and *in situ* bacteria. Of the two possibilities, the greatest obstacle to the cellulase action was probably the cooked starch. Cooked starch interacts with the cellulose molecules, protecting them from the enzyme. Interactions between *in situ* bacteria and enzymes were reportedly unusual, but not unprecedented in wastewaters with high bacterial counts (Bearzatto, 2006).

The reduction in COD when wastewater was acidified and treated with enzymes was high. However, the untreated refrigerated samples also experienced COD reduction. Therefore, the reduction was not due to enzyme treatment or microbial digestion. It might have been caused by acid hydrolysis or settling out of organics.

Biological digestion

None of the treatments tested gave satisfactory reduction in COD as well as increased filterability. Differences in results from various treatments were more obvious at longer retention times, so the retention times of one day were probably too short to give meaningful results. The addition of bacterial culture did not give significantly different results, possibly because the high levels of *in situ* bacteria masked any effects.

Although none of the treatments tested were successful, many other digestion treatments exist which could be tried. Moving bed bioreactors, for example, are reported to remove 60-65% of SCOD, and 90-95% carbohydrates in thermomechanical pulp wastewater (2500-3500 mg/L) with 13-22 hours retention time (Jahren *et al.* 2002).

Wet oxidation

Cooking time was the most significant factor contributing to successful wet oxidation. NaOH had no effect on COD reduction and decreased filterability, making it the least successful treatment. Using acetic acid achieved some very high reductions in COD (up to 100%) but unfortunately it also decreased filterability and corroded some of the copper in the reactor. Lower doses or a different acid could be investigated. The combination of H_2O_2 with reagents gave the good results, although without reagents H_2O_2 appeared to have no effect. The change to the reactor design after the initial test run to improve mixing of air and steam may have contributed to inconsistencies in the results. It is uncertain if inorganic species in the reactor from corrosion products (Fe, Cu, Mg^{+2}) were having a catalytic effect or affecting the reaction in any other way. Further investigation would be needed before this process could be applied to other wastewaters.

Conclusions

Amcor may wish to use these results to guide further exploration of treatment methods. The best treatment using the criteria of COD reduction and increased filterability was wet air oxidation. Further testing is needed to determine optimum operating conditions. The wastewater is unsuitable for enzyme treatment. The aerobic digestion treatment tested did not improve wastewater quality within a suitable time frame. However, other digestion treatments could be tested which might achieve higher efficiencies.

Acknowledgments

This project was supported by: the Western Australian Office of Science and Innovation, Amcor Spearwood Mill (General Manager Mr Ming Leung, Environment & Technical Development Manager Charles Bucknell and other staff), the suppliers of the biological treatments (Enzyme solutions, World Enzymes, and Eco-green) and Shayne Regan from Ciba.

References

Bearzatto, G. (2006). Personal communiation- telephone call. 23rd February 2006.

Clark, J.A., 1985. *Pulp Technology and Treatment for Paper*. Second Edition. Miller Freeman Publications, San Francisco, USA.

Dalentoft, E., and Thulin, P., 1997. The use of Kaldness suspended carrier process in treatment of wastewaters from the forest industry. *Wat. Sci. Tech.*, 35(2-3) pp123- 130.

Habets, L. H.A., and Knelissen, H.J., 1997. In line biological water regeneration in a zero discharge recycle paper mill. *Water Science and Technology*, 35, (2-3), pp 41-48.

Jahren, S.J., Rintala, J.A., and Odegaard, H., 2002. Aerobic moving bed biofilm reactor treating thermomechanical pulping wastewater under thermophilic conditions. *Water Research* 36 (4): 1067-1075 FEB 2002

Karam, J., and Nicell, J.A, 1997. Potential Applications of Enzymes in Waste Treatment. *J. Chem. Tech. Biotechnol.* 69, p 141- 153.

Minowa, T., & Inoue, S., 1998 .Hydrogen production from biomass by catalytic gasification in hot compressed water

Motherwell, G., 2006. Note . Personal communication 25th January 2006.

Smook, G.A., 1988. *Handbook for pulp and paper technologists*. Joint Textbook Committee of the Paper Industry. Canada.

Tardif, O., and Hall. E.R., 1997. Alternatives for treating recirculated newsprint wastewater at high temperatures. *Water Science and Technology,* 35(2-3) pp 57-65.

Verenich, S., Laari, A., Nissen, M., and Kallas, J., (2001). Combination of coagulation and catalytic wet oxidation for the treatment of pulp and paper mill effluents. *Water Science and Technology*. 44(5) p145-152.

12.4 Comparison of Ozonation and Photooxidation of Phenol

Patcharaporn Suwanvitaya[1], Panumas Puongkaew[2], Kitiyos Tungsudjawong[2]

[1]Associate Professor, Dept. of Environmental Engineering, Kasetsart University
[2]Dept. of Environmental Engineering, Kasetsart University

Abstract

Generally, aromatic compounds are toxic organic compounds. Among them, phenol which is used in various industries, has been found to contaminate the environment, especially in wastewater. Owing to toxicity and low biodegradability of phenol, advanced oxidation process is an alternative of interest. It can destroy aromatic ring structure of phenol, resulting in better biodegradability. This study compared two AOPs in the oxidation of phenol; ozonation and photooxidation with H_2O_2 . In the batch experiment, the parameters varied were; pH, phenol concentration (25 to 500 mg/l), ozone dosage and phenol to H_2O_2 molar ratio.

The results indicated that ozonation was effective in the lower range. The ozonation at 34 mg O3/l was effective in phenol destruction in the low phenol concentration range. Photooxidation, on the other hand, was more effective in the higher concentration range, with suitable phenol to H_2O_2 molar ratio. Although phenol was significantly reduced, the total organic carbon was still high, mainly in the form of organic acid. Results of the determination of organic acid by means of HPLC were also presented.

Key Words

Ozonation, UV/H_2O_2, phenol, advanced oxidation process

Introduction

Phenol is one of aromatic compounds widely used in various types of industries. When contaminating the environment, it can be toxic to human, animals, plants and microorganisms. The effects can be acute or chronic. Direct contact to skin can causes irritation. Inhaling can damage tissues in respiratory system. Accumulation of phenol in a body for long period may cause cancer.

In Thailand, phenol is used as a precursor or admixture in many industries, including plastic, petrochemicals, agrochemicals, cosmetics and pharmaceutical industries. It is commonly found contaminating wastewater in varying amount depending on production processes.

While benzene ring structure of phenol is not ready for biological degradation, toxic substance can be harmful to microorganisms responsible for the degradation. This can lead to a problem in wastewater treatment system. Advanced oxidation process (AOP) thus is an alternative method of treatment. Oxidation of low or non-biodegradable substances by AOPs achieved with the use of strong oxidizing agent, primarily hydroxyl radical (OH°). The processes include :- Ozone, Ozone / H_2O_2, Ozone / UV, UV / H_2O_2 or UV / H_2O_2/Ozone. They can alter complex (non-biodegradable molecule) to simpler substance, make it biodegradable. Toxic substance may also be destroyed, thus its toxicity reduced. [1] [2]

Ozonation : ozone is one of the strong oxidizing agent capable for oxidation of various types of substances including aromatic hydrocarbon. Complex molecule of some organic compounds can be degraded to simpler structure render it to be more biodegradable [3] [4]. Oxidizing agents may form from the reaction of ozone, making the process even more efficient [5]. The process can be only an ozonation, or a combination of ozone and the other oxidation process. The simplest and most practical methods of ozone production is the electrical (corona) discharge. Air is passed through high voltage region, oxygen in the air is ionized and then associated with non-ionized oxygen. Formation of ozone is as follow.

$$O_2 + e^- \rightarrow 2(O) + HEAT \qquad (1)$$

$$2(O) + 2O_2 \rightarrow 2O_3 \qquad (2)$$

Photo-oxidation : OH° can be generated by photochemical means either by UV radiation (100-400 nm.) or visible radiation. The wavelengths is determined by the principles of OH° production. The process most commonly used is generation of OH° through photolysis of conventional oxidant, H_2O_2. The following equation described photolysis of H_2O_2.

$$H_2O_2 + \text{light energy} \rightarrow 2OH^{\circ} \qquad (3)$$

In this study, phenol was treated in ozonation and UV/ H_2O_2 batch reactors. The experiments were conducted under varied conditions of pH, initial phenol concentration, ozone dosage, and amount of H_2O_2. Oxidation of phenol by the 2 processes were compared. The products from the processes were also determined.

Materials and Methods

Reactor

Ozonation reactor was an acrylic reactor, 10cm. in diameter x 100 cm. high. Ozone generator was of a household type producing 1.14 mg.O_3/min. Ozone was dosed from the bottom part, dosage was regulated by generator running time. Tube from the upper part led off-gas to KI solution, for ozone determination by Wet Chemistry Potassium Iodide methods [6]. The diagram of ozonation reactor is as shown in figure 12.4.1.

UV/H_2O_2 photolysis was conducted in stainless steel reactor of the same size. UV lamp emitting 315 nm. wavelength was in the inner ring with water sample in the outer ring. Figure 12.4.2 shows diagram of photolysis reactor.

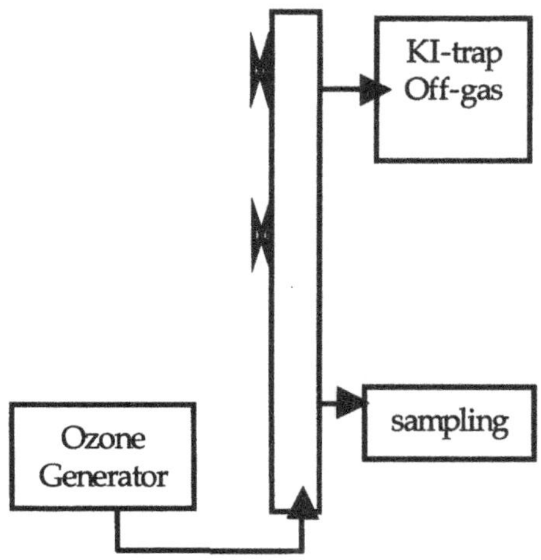

Figure 12.4.1: diagram of ozonation reactor

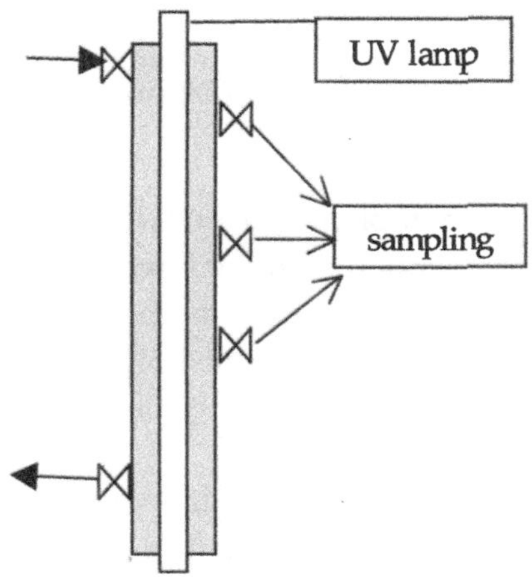

Figure 12.4.2: diagram of UV/H_2O_2 reactor

Experimental procedure

Water sample was prepared from granular phenol to contain 25, 50, 100, 150, 250 and 500 mg. phenol/L. pH of the samples was adjusted to 4, 7, and 10 with H_2SO_4 and NaOH. Ozonation was conducted under 9.5-34 mg./L range. For UV/H_2O_2 photolysis, H_2O_2 was applied at the H_2O_2 : phenol molar ratio of 50:1, 150:1 and 300:1. Contact time was varied from 0 to 140 min.

Analytical procedure

Parameters investigated were phenol and Total Organic Carbon and volatile fatty acid. Phenol removal was determined from the remaining amount of phenol. Only some portions of phenol could be completely oxidized, products of incomplete oxidation would remain in the form of organic compounds which could be determined by TOC concentration. Results from previous studies showed that the breaking down of benzene ring structure yielded some volatile fatty acids [7]. VFA was thus analysed. Analytical methods used for the above mentioned parameter were as shown in table 12.4.1.

Table 12.4.1: analytical methods for studied parameters

Parameter	Analyses / instrument
Phenol	4-aminoantipyrine, measure absorbance at 500nm. Spectrophotometer (HACH – DR 4000)
TOC	Total Organic Carbon – SHIMADSU 5000A
Fatty acids	HPLC (AGILENT 1100), VFA Column 100 % Phosphate buffer pH 2.5, Mobile Phase flow rate 1 mL/min

Results

Phenol oxidation by ozonation.

Ozonation of phenol

For ozonation of phenol, ozone directly reacted with phenol. As can be seen in table 12.4.2, amount of phenol remaining after treatment decreased with an increase in ozone dosage. However, with the production capacity of the generator used in this study, the amount of ozone generated was capable only for the lowest phenol concentration of 25 mg./L. for the sample with higher amount (50 and 100 mg./L) only small fraction of phenol was removed.

Considering the effect of pH, it could be noticed that with the same concentration of phenol and ozone, pH variation significantly affected phenol removal. More phenol could be removed at pH 10.

Table 12.4.2: phenol remaining (mg/L) under varying condition of ozonation

Initial concentration mg.phenol/L		Ozone dosage (mg/L) 0	9.5	17	25.5	34
25	pH 4	24.91	18.43	13.61	9.98	6.23
	pH 7	26.53	18.81	12.84	8.11	5.72
	pH 10	24.05	13.48	7.21	2.57	0.69
50	pH 4	45.24	41.92	37.61	33.52	28.10
	pH 7	46.95	38.5	33.94	28.44	23.20
	pH 10	46.56	37.78	29.81	20.98	15.87
100	pH 4	97.65	93.81	86.67	79.10	73.67
	pH 7	97.11	86.99	77.29	72.10	66.21
	pH 10	97.43	85.92	74.52	63.01	54.59

UV/H_2O_2 photolysis of phenol

For UV/H_2O_2 photolysis, main oxidant was OH°, induced from UV radiation H_2O_2. Thus UV exposure time was also played a role in phenol oxidation in the process with sufficient amount of H_2O_2. Table 12.4.3 showed the results obtained from oxidation of phenol in the present of 150:1 H_2O_2 : phenol molar ratio under varying pH condition and exposure times. It was found that amount of phenol removed increased with an increasing in exposure time. The process was capable to remove higher amount of phenol than ozonation.

pH seemed to have no significant effect on UV/H_2O_2 photolysis. Varying pH under the same conditions resulted in slightly different removed amount.

Table 12.4.3: phenol remaining (mg/L) by UV/H_2O_2 with 150:1 H_2O_2: phenol molar ratio

Initial concentration mg.phenol/L		UV Exposure time (min) 0	5	10	15	20	30
25	pH 4	48.44	10.13	2.8	0.67	0.1	0.06
	pH 7	44.00	5.85	1.28	0.23	0.14	0.08
	pH 10	45.1	17.1	3.55	0.72	0.13	0.09
150	pH 4	143.1	64.1	25.3	7.92	0.59	0.04
	pH 7	147	44.4	23.8	7.5	0.97	0.01
	pH 10	148	76.5	19.7	6.1	0.86	0

As previously mentioned that phenol oxidation photolysis depended on OH° production, phenol concentration was then closely related to the amount of H_2O_2 and UV exposure time. The results in table 12.4.4 showed the effect of H_2O_2 amount and UV exposure time on oxidation of various concentration of phenol at pH7. These 2 components had a combination effect. It was clearly seen that phenol was removed more rapidly with an increase in H_2O_2 : phenol molar ratio. However, with higher initial phenol concentration, longer exposure time or higher molar ratio was needed to achieved removal at the same level, for example 30 min. with 10:1 for 50mg.phenol/L., 60 min. with 50:1 for 150mg.L., 90 min. with 50:1 for 250mg.L. and 120 min. with 50:1 for 500mg.L.

Table 12.4.4: Residual phenol remaining (mg/L) by UV/H_2O_2 at pH7, under varying exposure time and H_2O_2: phenol molar ratio

Phenol mg./L	m.H_2O_2: m.phenol	UV Exposure time (min) 0	5	10	20	30	60	90	120
50	10:1	51.77	14.68	5.77	1.34	0.10			
	50:1	49.7	6.45	2.59	0.34	0.19			
	150:1	44	5.85	1.28	0.14	0.08			
	300:1	49.2	14.4	6.1	0.34	0			
150	10:1	153.34	65.27	38.84	24.56	7.29			
	50:1	149.3	49.3	20.3	8.78	0			
	150:1	147	44.4	23.8	7.5	0.01			
	300:1	152.5	44.1	17.84	5.13	0			
250	50:1	234	128.4	85.99	50.77	24.56	5.58	0	
	150:1	262	162	99.7	44.2	25.9	1.91	0	
	300:1	231	101.8	56.76	31.06	13.79	0.74	0	
500	50:1	504.3	272.9	236.5	204.3	168.7	97.7	35.5	0
	150:1	491.5	283.6	239.5	185.5	132.9	45.5	0	0
	300:1	492.6	246.1	203.2	149.9	104.3	14.64	0	0

Oxidation products.

Results of TOC analyses included all organic compounds in the form of weight of carbon The decrease in organic carbon weight (TOC) was the portion that underwent complete oxidation and escaped out in the form of carbon dioxide gas. Figures 12.4.2 and 12.4.3 show the amount of removed phenol calculated in terms of carbon (phenol carbon)was compared to the removed TOC. With both oxidation process, significant amount of phenol was destroyed, while only some portions of organic carbon was removed, indicating that only some portions of phenol underwent complete oxidation.

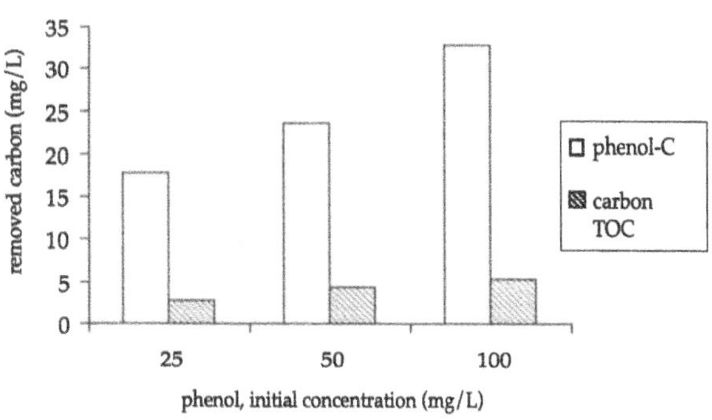

Figure 12.4.2: amount of phenol-carbon and TOC removed by ozonation

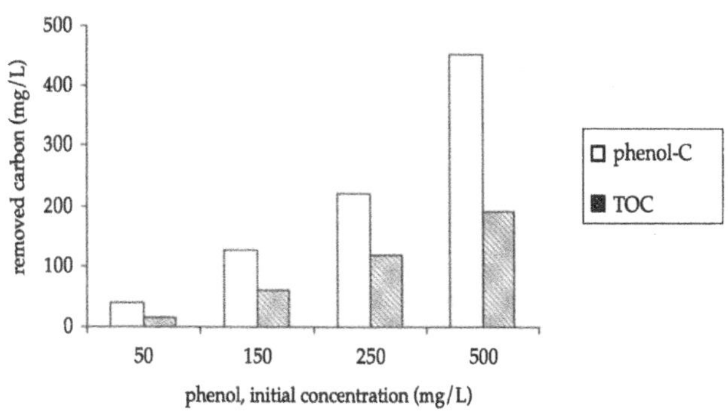

Figure 12.4.3: amount of phenol-carbon and TOC removed by UV/ H_2O_2 photolysis

It is clear from the comparison of figures 12.4.2 and 12.4.3 that for ozonation process, TOC was very low, compared to the phenol carbon. For the UV/ H_2O_2 photolysis, this was much higher. This indicated that higher proportion of TOC was removed by UV/ H_2O_2 photolysis than by ozonation.

Treated samples were further analysed for organic acid content by HPLC techniques (formic acid, acetic acid, propionic acid and oxalic acid). It was found that treated samples contained all the four acids to a varying amount plus a significant portion of unidentified organic acid. This confirmed that not all phenol underwent complete oxidation but only partially oxidized. Some of the oxidation products were volatile organic acids.

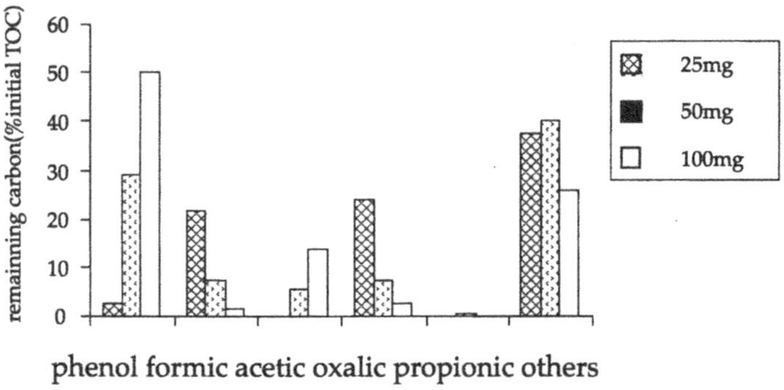

Figure 12.4.4: Phenol and volatile fatty acid content in samples treated by ozonation.

Conclusion

1. Ozonation was pH dependent, with pH10 yielding better performance. pH level did not significantly affect the performance of UV/ H_2O_2.
2. UV/ H_2O_2 had combined effect on phenol oxidation. With higher phenol concentration, either longer exposure time or larger amount of H_2O_2 was needed to achieve the same removal level.
3. Not all phenol destroyed underwent complete oxidation. Volatile fatty acids (formic, acetic, propionic and oxalic acids) were found in the treated samples, indicating the formation of incomplete oxidation products.

Reference

Santiago Esplugas, Jaime Giménez, Sandra Contreras, Esther Pascual and Miguel Rodríguez, 2002. Comparison of different advanced oxidation processes for phenol degradation, *Water research*, 36 (4) : 1034-1042

Mokrini,A., D. Ousse and S. Esplugas, Oxidation of aromatic compound with UV radiation / ozone / hydrogen peroxide. *Wat.Sci.Tech.*, 38(4):95-102.

Baig, S., I. Coulomb, P. Courant and P. Liechti, 1999. Treatment of landfill leachates: Lapeyrouse and Satrod case studies, *Ozone Sci. Eng.*, 21 : 1-22.

Bila, D.M., F. Montalvao, A.C. Silva and M. Dezotti, 2005. Ozonaton of a landfill leachate : evaluation of toxicity removal and biodegradability improvement, *J. Hazard Meaterial*, 117(2/3) : 235-242.

Chang, D.S. and P.C. Singer, 1991. The impact of ozonation on particle stability and the removal of TOC and THM precursors. *J. AWWA*, 83(3):71-78.

Lenore S. Clescerl, Arnold E. Greenberg and *Andrew D. Eaton,* 1999. Standard Methods for the Examination of Water and Wastewater ,20th. USA : AWWA.

Gottschalk, C., J. A. Libra and A. Saupe. 2000. *Ozonation of Water and Waste Water.* Germany : WILEY-VCH, Inc.

13 Rainwater Harvesting

13.1 An Innovative Rainwater Management Regulation in Seoul Metropolitan Government, Korea

Mooyoung Han, Professor, Seoul National University (myhan@snu.ac.kr)

Woongki Song, Director of water quality division, Seoul Metropolitan Government (songcity@seoul.go.kr)

Abstract

A new paradigm for rainwater management is suggested as a means of reducing water problems that arise from urban flooding and drought because of climate change in Korea. The rainwater management system uses many decentralized rainwater tanks for more than one rainwater management purpose. Multipurpose rainwater management is made possible by the use of information and control technology together with the cooperation of citizens. The system has been implemented by regulation in Seoul City.

Key Words

Rainwater management, Multipurpose, Information technology, Climate change, Korea

Introduction

Korea has always suffered from frequent flooding and drought because it is located in a monsoon area. Flooding and drought have increased in severity, probably because of climate change coupled with increased surface runoff caused by urbanization. This has resulted in many casualties and loss of property, as well as the high cost of securing sewerage and river systems.

The rainfall characteristic of Korea is compared with that of other countries. Figure 13.1.1 represents the comparison of both the average and variance of annual rainfall. The average annual rainfall of Korea is relatively high (1,283 mm) and comparable with other countries such as Brazil, Japan, and Argentina. However, when comparing the variance of annual rainfall, Korea (12,000mm^2) has much higher value than any other country in the world like Japan(2,000 ~ 4,000), Brazil(4,000 ~ 6,000) and Argentina (below 2,000). This means, in Korea, much of rainfall is in summer season while less rainfall is in winter season. There are not many countries in the world that suffer both flooding and drought problem at the same time as is the case in Korea. Therefore, it is natural that there should be a unique method to survive in the severest rainfall characteristic of this country.

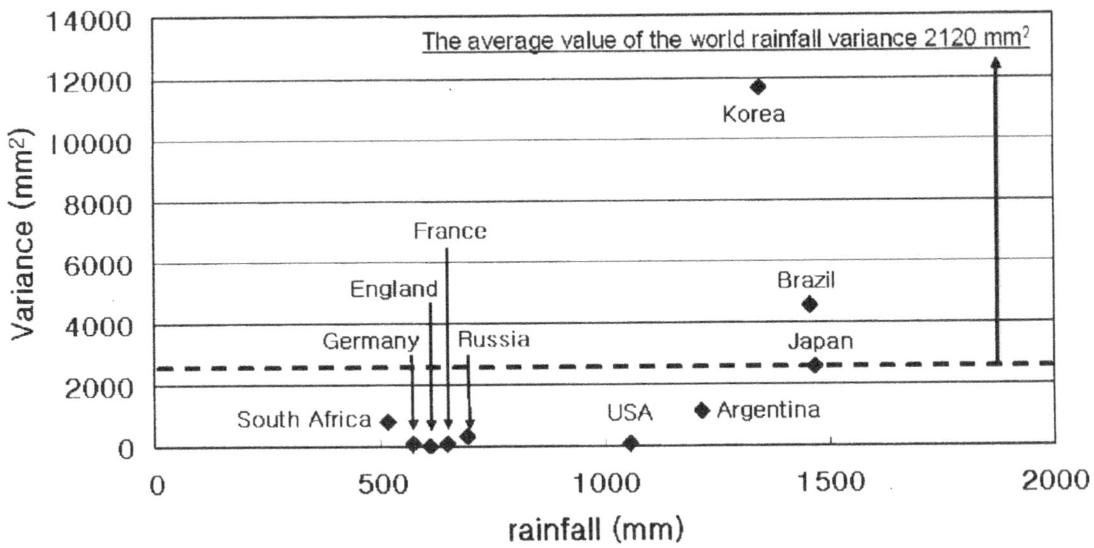

Figure 13.1.1: Rainfall – variance of each country

A new paradigm is suggested where rainwater is managed on a decentralized basis, is controlled near its source, and involves local activities. The concept of multipurpose rainwater management is suggested for a more efficient management system.

Three examples of multipurpose rainwater management in Korea are discussed. The first example is a design concept for a rainwater tank in a new building in the Star City project. The tank is divided into three compartments: one for flood prevention, another for water conservation, and the third for emergency use.

The second example is a new regulation made by the Seoul Metropolitan Government, which requires installation of rainwater tanks in new large buildings for the purposes of flood prevention and water conservation. The disaster prevention board of Seoul City can control the multiple rainwater tanks by using information and control technology. When heavy rainfall is expected, the City can order building owners to empty the tanks to prevent sewer flooding. At other times, the stored rainwater can be used for non-drinking purposes as a water conservation measure.

Cities elsewhere that have unexpected water pollution could use the multipurpose rainwater management concept suggested in Korea.

The third example is the regulation of Kwangjin-Gu, Seoul. An incentive program is initiated that allow the building owner to build more space area according to the extent of rainwater system at Kwangjin-Gu, Seoul.

Changing rainfall characteristics due to climate change

Seoul City is 600 years old and has a population of 10 million. Comparison of the rainfall averages over the last five and 30 years shows the effect of climate change (Figure 13.1.2). Overall precipitation has increased with more rainfall in summer and less in spring. As a result, many casualties and much loss of property occur each year. A very large investment is being planned to prevent flood damage.

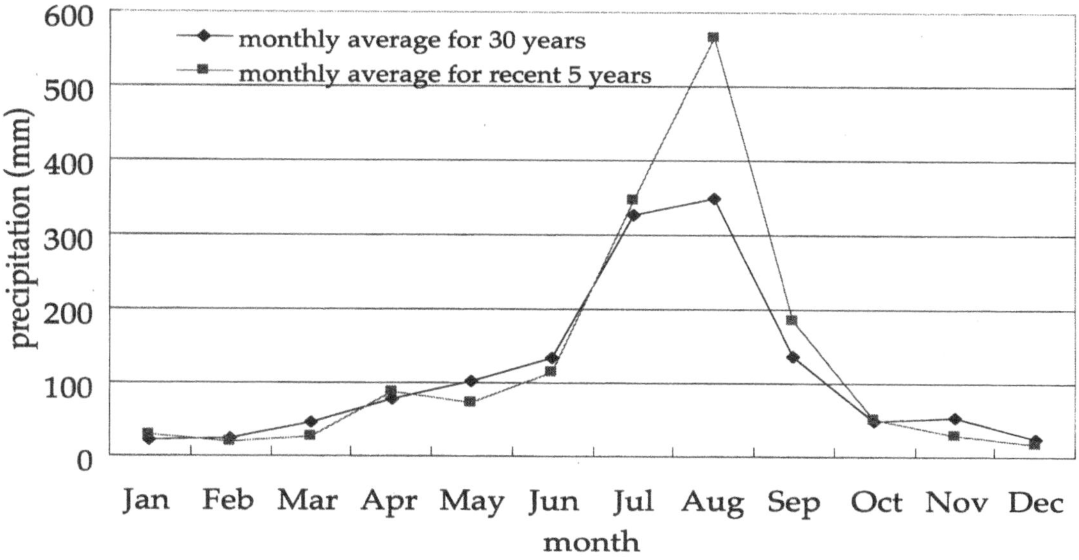

Figure 13.1.2: Change in average monthly rainfall in Seoul

The question arises as to whether the capacity of the existing water conveying system, such as sewers and rivers, is sufficient and, if not, how to increase the capacity.

Traditional thinking has been to upsize sewers, widen rivers, or increase pumping capacity when carrying capacity becomes inadequate. The water conveying system could be reaching its limit because of the recent rainfall characteristics. The cost of upgrading is very high and there are space restrictions on constructing new facilities on or under roads in a congested city.

Hydraulic analysis of a circular sewer pipe shows that, when flow is near its peak, flow capacity decreases suddenly after a small addition of water. This causes sudden urban flooding. Therefore, when a sewer pipe is close to its full capacity, a small volume of water diverted to a water tank can reduce the danger of urban flooding at a local level.

A new paradigm for rainwater management

The traditional approach to urban flooding is to convey rainfall over a wide area to sewers and rivers.

An alternative concept is to manage rainfall in its watershed; that is, to store and manage the rainwater in the local area. By storing rainwater in a large number of small tanks, peak runoff is reduced and the time to reach peak flow is delayed. Relatively clean water stored in the tanks can be used for non-drinking purposes, resulting in water conservation. However, this approach requires the understanding and cooperation of local people, who must build many tanks and operate them according to rules.

A new paradigm is to develop multi-purpose rainwater management facilities. For example, it would be more efficient to manage both flooding and drought simultaneously.

Concept of multipurpose rainwater management

There are many water problems in Korea, such as flooding, drought, water pollution, dry rivers, and mountain fires. As briefly discussed below, these problems remain, possibly because each problem has been dealt with separately in the past.

Flood mitigation

Large structures, such as dams, and riverbank raising, are considered for flood prevention. However, they cannot prevent flooding in local areas and small-scale sewer inundation occurs. These structures are not used to their full capacity most of the time.

Drought management

Groundwater extracted in spring, when there is no rainfall for an extended period, is being rapidly depleted. Water quality problems arise in rivers because there is insufficient water for dilution.

Dry river restoration

Some rivers in the city become dry during the dry season because of the lowered groundwater level caused by slow infiltration of rainwater and by fast runoff during rainfall from covered impermeable surfaces. Measures to pump treated wastewater effluent or downstream river water upstream require large amounts of energy and result in water quality problems.

Mountain fires

Mountain fires, which can be controlled by water, occur frequently in spring. However, the firefighting agency has little control of water availability. Unfortunately, mountain fires burnt a 1,000-year-old temple and its treasures in March 2005.

Individual ministries with different interests and priorities deal with water problems. Solutions may sometimes be shortsighted and inefficient in national terms. The simple rose diagram shown in Figure 13.1.3 illustrates this approach. Each law is shown conceptually. The river law, drawn in red, shows that the major interest is flooding, with very little concern for drought. The water supply law deals only with water resource management, with little emphasis on other major fields.

An ideal management model would encompass all major items. However, the current major concerns in Korea are flooding, drought, and fire prevention.

The Seoul Metropolitan Government is now enforcing the first example of multipurpose rainwater management. Seoul has been suffering from repeated damage by flooding because it is urbanized and covered with impervious surfaces.

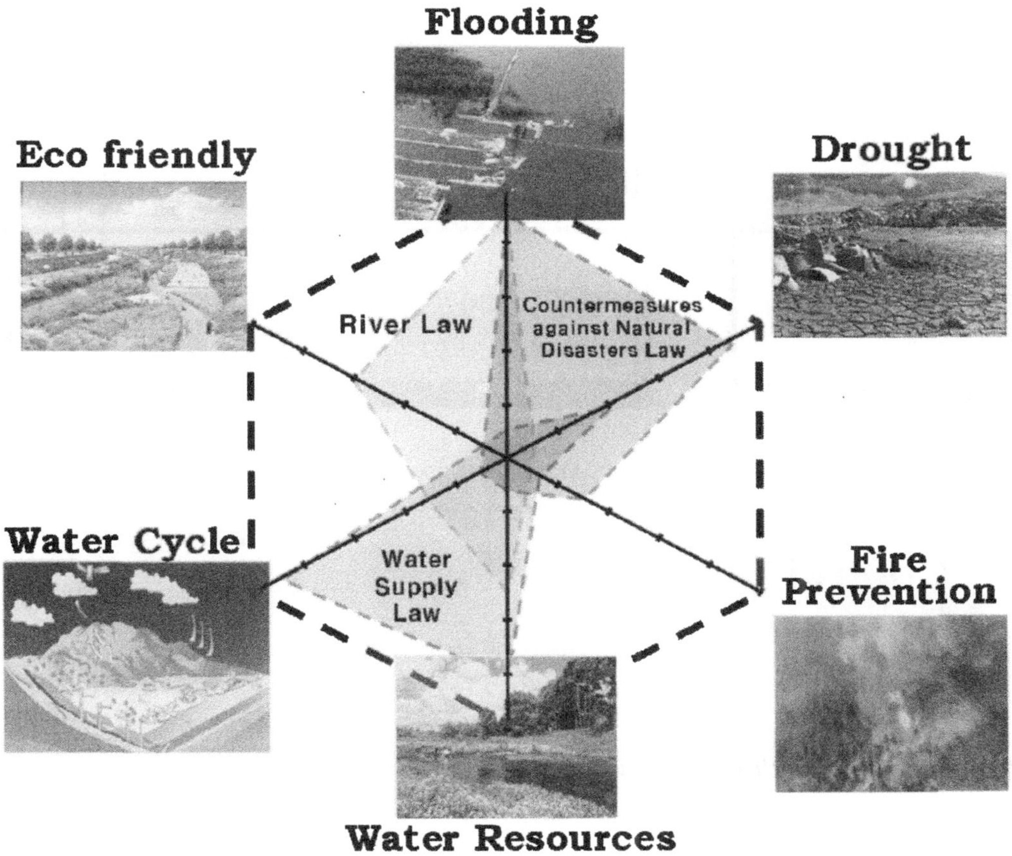

Figure 13.1.3: Concept of multipurpose rainwater management

Examples of multipurpose rainwater management

Design of a multipurpose rainwater tank in a building project

A specific rainwater system was designed for a recently constructed building (Figure 13.1.4) at the Star City Project in Kwangjin-Gu, Seoul. A 3,000 m^3 rainwater tank was installed in the basement and divided into three sections of 1,000 m^3 each. The first section collects rainwater from the unpaved surface. It should be kept empty most of the time except when there is heavy rain. The second 1,000 m^3 section collects rainwater from the roof, which should be used for toilet flushing and landscaping purposes. The third 1,000 m^3 section should be filled with fresh water and used for supply during emergencies such as firefighting or accidents.

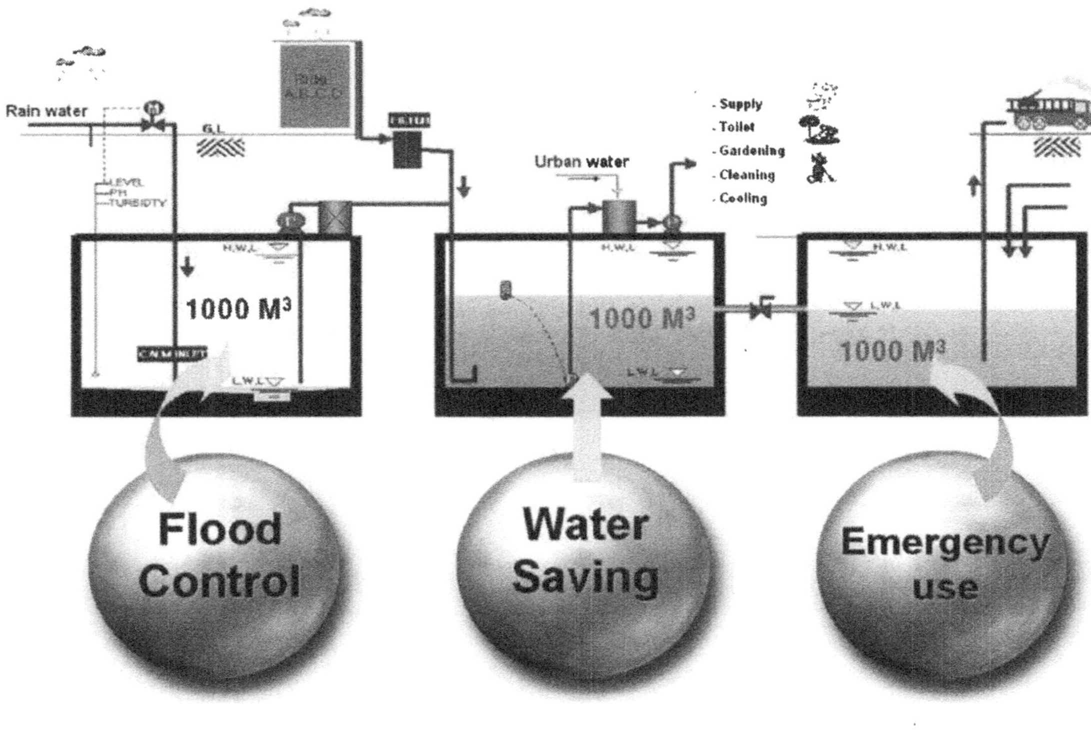

Figure 13.1.4: Example of multipurpose rainwater tank design

Seoul City rainwater management

Seoul City announced a new regulation to enforce the installation of a rainwater harvesting system in December 2004. The main purpose is to mitigate urban water flooding. The secondary purpose is to conserve water. It expects to ensure the safety of the city and improve the well being of citizens as a result. It asks citizens to voluntarily cooperate by filling and emptying the rainwater tanks according to directions from the disaster prevention agency.

A special feature of the new system is to provide a network for monitoring water levels of all water tanks at the central disaster prevention agency (Figure 13.1.5). Depending on the expected rainfall, the central disaster prevention agency may issue an order to building owners to empty their rainwater tanks fully or partially. An incentive program is planned for those who follow the order and some punishment for those who do not. After a storm event, the stored water can be used for firefighting and/or miscellaneous purposes such as toilet flushing and gardening.

The buildings included in the regulation are:

- All public buildings: compulsory for new buildings and recommended to the extent possible for existing buildings.
- New public facilities such as parks, parking lots, and schools: to the extent possible.
- Private buildings: recommended for new buildings subject to building permission (floor area larger than 3,000 m^2).
- Large development plans such as new town projects: installation of a rainwater management system as a first priority.

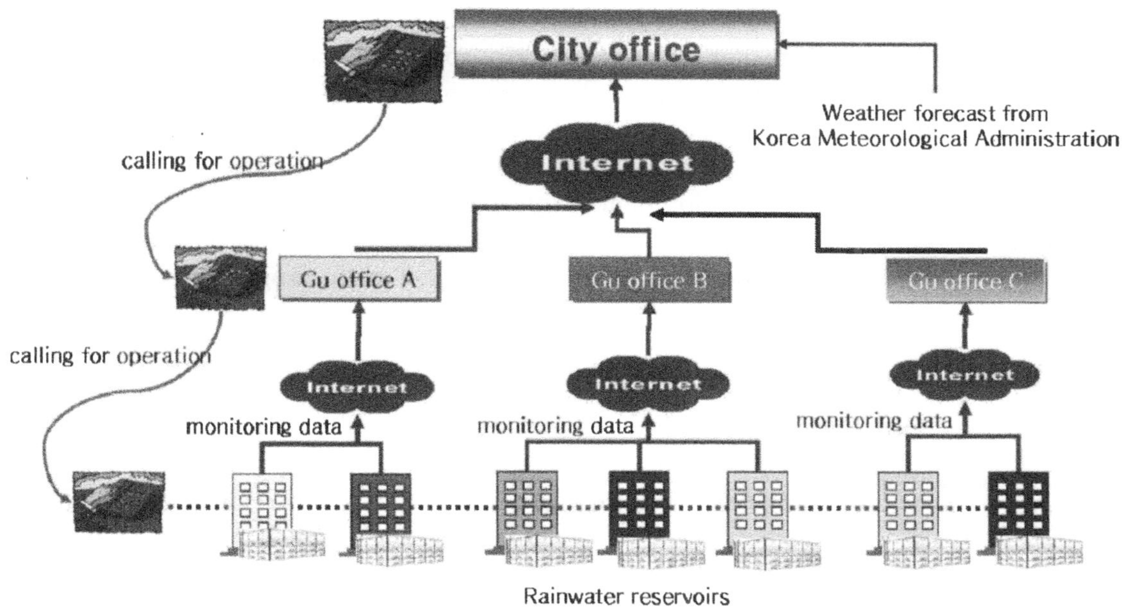

Figure 13.1.5: Diagram showing the monitoring of a multiple tank system for preventing urban flooding and for water conservation

Incentive program in Kwanjin-Gu, Seoul

An incentive program is developed for the installation of a rainwater harvesting system in Kwanjin-Gu, Seoul. Special merits of the new incentive program which allow the building owner to build more floor area are remarkably realizable. The incentive program with deliberations of the Urban, Architecture committee is follows:

Floor area incentive program (Kwangjin-Gu)

- A building which has rainwater system only for the purpose of flooding prevention is allowed for 5% more floor area
- A building which has rainwater system for outdoor water use such as for gardening, washing and flushing is allowed for 10% more floor area
- A building which has rainwater system for Indoor water use such as for flushing and washing floor is allowed for 15% more floor area
- A building which has rainwater system for both outdoor and indoor water use is allowed for maximum 20% more floor area

Conclusion

Damage by flooding is increasing in Korea because of climate change. A new paradigm for rainwater management is suggested to reduce the cost of flood mitigation. The concept is to use many decentralized rainwater tanks as a multipurpose rainwater management system. The examples of multipurpose rainwater management at a building project and by regulation in Seoul City to prevent flooding and improve water conservation are promising. Operation and control of multiple tanks in conjunction with weather information is possible with the development of information and control technology and with the understanding and cooperation of citizens. Other innovative ideas and projects based on the concept of multipurpose rainwater management are in progress.

References

http://rainwater.snu.ac.kr/

http://www.seoul.go.kr/

13.2 Insight from a Decade of Research into the Quality of Rainwater Supplies Collected from Roofs

Peter J Coombes

School of Environment and Life Sciences, University of Newcastle
Callaghan, 2308, Australia
p.coombes@newcastle.edu.au

Abstract

A research program into the quality of water supply from rainwater tanks over the last decade at the University of Newcastle in Australia has provided insights into water quality processes operating in rainwater tanks. A rainwater treatment train has been identified that includes processes in first flush devices, rainwater tanks and hot water services including the action of biofilms and heat death of bacteria in hot water services. The need for continuing scientific endeavour to improve understanding of this important water source is highlighted.

Introduction

Over 3 million Australians, in mostly rural areas, rely on rainwater tanks for drinking water supplies. It was a paradox that, prior to the 1990s, the use of rainwater tanks was virtually illegal in urban areas that had reticulated (mains) water supplies (Lloyd *et al.*, 1992). It is commonly claimed by authorities that rainwater tanks were removed from urban areas due to health concerns. In contrast, Lloyd *et al.* (1992) explain that rainwater tanks were removed to increase the economic viability of water utilities. No health epidemics have resulted from use of rainwater. Experience has shown that health risks resulting from the consumption of rainwater are low throughout Australia [Cunliffe, 2004].

Research has revealed that the use of rainwater tanks in urban areas to supplement rainwater supplies can significantly improve the security of regional water supplies, and reduce the stormwater quality and quantity impacts of urban development (Coombes, 2006; Coombes *et al.*, 2002). Recently, in response to water shortages, authorities throughout Australia encourage the use of rainwater tanks in urban areas to supplement mains water supplies and to manage urban stormwater runoff. Although rainwater tanks are widely used in Australia there is limited understanding about the quality of rainwater supplies. This paper provides an overview of a decade of research into the quality of rainwater conducted at the University of Newcastle in Australia.

Monitoring Sites

Monitoring of demonstration projects has provided valuable insight into the performance of rainwater harvesting systems. These projects include the Figtree Place, Maryville and Carrington sites that are presented below.

Figtree Place

Figtree Place is located in Hamilton an inner city suburb of Newcastle in New South Wales Australia. The site was developed as a prototype site for water sensitive urban design that includes 27 residential units, rainwater tanks, infiltration trenches and a central basin where cleansed stormwater enters an unconfined aquifer for water retention and retrieval (Figure 13.2.1).

Rainwater runoff from roofs at Figtree Place passes through first flush pits and is captured in four underground tanks with capacities ranging from 9 kL to 15 kL for toilet flushing and hot water use in the residential units. Rainwater supplies are drawn from a depth of 0.4m from the base of the tanks. A low level monitor is used to transfer to mains water supply when water levels are low in the rainwater tanks (see Coombes *et al.*, 1999; 2000).

The Figtree place project was proposed in 1995. However construction was not completed until 1998 due to strong resistance from water authorities and government departments to the use of rainwater for toilet flushing and in hot water services. Over 3.2 million Australians rely on rainwater for drinking water without widespread health impacts. Nevertheless authorities claimed that consumption of rainwater posed a high hazard that was likely to cause death. A combination of institutional resistance and construction errors led to the installation of underground tanks with holes in walls allowing entry of debris to the tanks. The Figtree Place site was also subject to high levels of air pollution due to location adjacent to roads with high traffic loads, a bus station and heavy industry. Poor quality rainwater was expected.

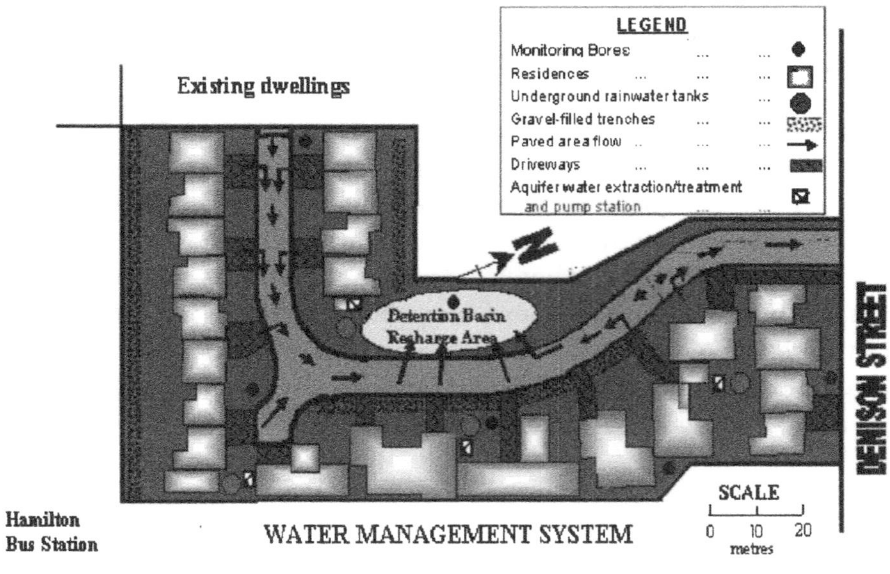

Figure 13.2.1: General plan and water sensitive elements of Figtree Place

A comprehensive monitoring program was implemented to understand water quality at the Figtree Place project (Coombes *et al.*, 1999; 2000). Samples collected from rainfall, roofs, at the water surface in tanks, at the point of supply in tanks and from hot water were tested for the range of parameters shown in Table 13.2.1. The water quality analyses were conducted in accordance with the Standard Methods for the Examination of Water and Wastewater (APHA, 1995). Annual average mains water savings of 45% were measured at Figtree Place.

The bacterial indicator organisms (Fecal and Total Coliforms) were measured for compliance with the Australian Drinking Water Guidelines. Heterotrophic plate counts can reveal the presence of organic material in water and are a general indicator of water quality. *Pseudomonas Spp*. can be used to indicate the presence of nutrients in water and of biofilms on surfaces. Average values for water quality at different locations in the rainwater supply system are compared to Australian Drinking Water Guidelines (Guideline) in Table 13.2.1.

Table 13.2.1: Average water quality parameters from the Figtree Place rainwater system

Parameter	Unit	Rain	Roof	Tank surface	Point of supply	Hot water	Guideline
Fecal Coliforms	CFU/ 100 ml	0	135	119	20	0	0
Total Coliforms	CFU/ 100 ml	0	359	834	166	0	0
Heterotrophic plate counts	CFU/ml	3	1,360	3,256	331	3	NA
Pseudomonas Spp.	CFU/ 100 ml	5,200	59,600	6,768	7,544	0	NA
Sodium	mg/L	9.9	11.4	4.9	5.4	4.4	180
Calcium	mg/L	2	2.7	6.9	8.4	10	200
pH	-	5.9	5.8	6.2	6.1	6.24	6.5 – 8.5
Dissolved solids	mg/L	21	57	98.2	114	94	500
Suspended solids	mg/L	8.4	35	1.4	1.6	0.78	500
Chloride	mg/L	7.5	14.9	7.1	10.5	10	250
Nitrate	mg/L	0.2	0.3	0.06	<0.05	<0.05	3
Nitrite	mg/L	0.4	2.2	0.6	1.0	0.8	50
Sulphate	mg/L	3.5	6.7	4.9	7.3	9.6	250
Ammonia	mg/L	0.3	0.4	0.1	0.3	0.18	0.5
Lead	mg/L	<0.01	0.014	<0.01	<0.01	<0.01	0.01
Iron	mg/L	<0.01	0.05	0.01	0.01	0.02	0.3

Table 13.2.1 demonstrates the general water quality processes in the rainwater supply system. The quality of rainwater was compliant with drinking water guidelines. However the quality of roof runoff exceeds the guidelines for Coliform bacteria and lead. Microbial and elemental contamination was observed to accumulate on roofs.

Concentrations of Total Coliforms, heterotrophic plate counts and dissolved solids increased at the water surface in the tanks in comparison to roof runoff. Concentrations of the remainder of the parameters decreased. The increases were attributed to entry of soil into the damaged underground tanks during rainfall events whilst the action of the first flush pits and settlement processes in the tanks were responsible for the decreased concentrations. The elemental quality of rainwater at the water surface was compliant with the drinking water guidelines.

At the point of supply in the tanks concentrations of heterotrophic plate counts, nitrite, Fecal and Total Coliforms were observed to decrease in comparison to concentrations at the water surface. The increased presence of Pseudomonas Spp. may indicate that biofilms have formed on the walls of tanks. Metabolic processes within biofilms can utilised nitrates and entrap bacteria (such as coliforms and heterotrophs) creating increases in ammonia. It was proposed that the formation of biofilms on tank surfaces contribute to improving water quality at the point of supply. The average concentrations of lead and iron remained low and compliant with the drinking water guidelines.

Concentrations of most of the remaining parameters increased slightly. This result suggests that metal and chemical parameters settle to the bottom of tanks where they sorb to sludge thereby improving water quality at the point of supply. Samples taken from sludge at the bottom of tanks show accumulation of lead (0.033 mg/L) and iron (0.093 mg/L) support the assumption that contaminants settle to the bottom of tanks. All parameters, except fecal and total coliforms, were compliant with drinking water guidelines.

Roof runoff captured in tanks is used to supply electric hot water storage services (capacity: 250 L) that are subject to "off peak" heating during the periods 6 am to 8 am and 8 pm to 10 pm. Although the microbial quality of tank water exceeded guidelines, the microbial quality of hot water was compliant with the drinking water guidelines. The processes of pasteurisation and tyndallisation (small perturbations in temperature) were believed to eliminate the majority of bacteria from the hot water. Monitoring and analysis at the Figtree Place site has revealed the presence of a rainwater treatment train that includes first flush device, the rainwater tank and the hot water service.

Maryville House

Using the lessons learnt from the Figtree Place project, an old house in Maryville, a neighbouring inner city suburb of Newcastle was fitted with a large above ground 9.1 kL Aquaplate™ rainwater tank to supply hot water, toilet and outdoor uses to the three person household (Figure 13.2.2) (Coombes *et al.*, 2003). An efficient dual water supply system was created that allowed the tank to be topped up with mains water to a minimum water level whenever rainwater supplies were low (Figure 13.2.3). This system eliminates the risk of backflow of rainwater into mains water supplies and ensures that water is always available. The average annual mains water savings provided by the tank was 52kL/annum.

Figure 13.2.2: The Maryville House

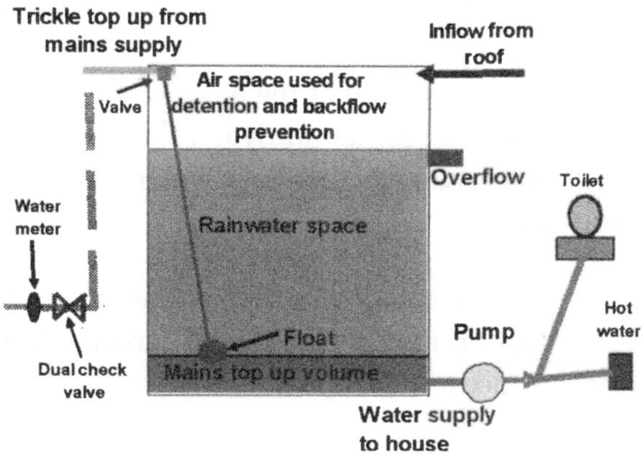

Figure 13.2.3: Design of the dual water supply system

Similar to the Figtree Place project, the Maryville House is located adjacent to Newcastle's heavy industrial area and the industrial highway. The site is subject to high levels of industrial and urban pollution. Unlike the Figtree Place project, the house has a rusty galvanised iron roof, an instantaneous gas hot water system that was set at 55°C and no first flush device. Average values for water quality at different locations in the rainwater treatment train are shown in Table 13.2.2.

Table 13.2.2: Average water quality parameters from the Figtree Place rainwater system

Parameter	Unit	Roof	Point of supply	Hot water	Guideline
Fecal Coliforms	CFU/100 ml	135	1	0	0
Total Coliforms	CFU/100 ml	359	18	0	0
Heterotrophic plate count	CFU/ml	1,360	784	4	NA
Pseudomonas Sp.	CFU/100 ml	59,600	1,673	0	NA
Sodium	Mg/L	11.4	7.5	8.4	180
Calcium	Mg/L	2.7	2.5	2	200
pH	-	5.8	5.7	5.5	6.5 – 8.5
Dissolved solids	Mg/L	57	67	16	500
Suspended solids	Mg/L	35	19	0.5	500
Chloride	Mg/L	14.9	9.9	9.9	250
Nitrate	Mg/L	0.3	<0.05	<0.05	3
Nitrite	Mg/L	2.2	1.4	1	50
Sulphate	Mg/L	6.7	5.9	5.2	250
Ammonia	Mg/L	0.4	0.3	0.2	0.5
Lead	Mg/L	0.01	<0.01	<0.05	0.01
Iron	Mg/L	0.05	<0.05	<0.05	0.3
Zinc	Mg/L	-	3.9	3.9	3

Table 13.2.2 reveals that the quality of rainwater at the point of supply in the above ground rainwater tank at Maryville was better than the quality at the point of supply in the damaged underground rainwater tanks at Figtree Place for most parameters.
The microbial quality of rainwater exceeded the guidelines for presence of Fecal Coliforms on a single occasion immediately following a rainfall event that preceded construction activity on the house roof. Notably, no fecal contamination was evident on the roof.

Fairly high levels of zinc were found in the rainwater that originated from the rusty roof. This level of zinc contamination was not considered a health hazard (NHMRC, 1996). Nevertheless, in spite of the absence of a first flush device, the quality of rainwater was seen to improve in the rainwater tank and the hot water service. A majority of parameters were compliant with drinking water guidelines. The rainwater treatment train was confirmed.

Elimination of bacteria in the instantaneous hot water service at Maryville indicated that pasteurisation may not be the dominant process for reduction of bacteria in domestic hot water services. It was postulated that the instantaneous heat differential between the rainwater tank and the hot water service also eliminated bacteria. Prescott *et al.*, (1999) report that moderate heat eliminates bacteria more readily at low population numbers, in acid conditions that are consistent with rainwater and with rapid changes in temperature.

Carrington House

A third rainwater harvesting system was installed at Carrington, an inner city suburb located in the docklands within the industrial area of Newcastle. Two small rainwater tanks (2.1 kL each) with a mains water top up scheme are used to supply all water uses to the two person household (Figure 13.2.4). The mechanical top up system relies on vertical movement of the float valve to top up the tank to a minimum level at a variable rate dependent on the water depth in the tank (Figure 13.2.5).

Figure 13.2.4: Picture of rainwater tanks

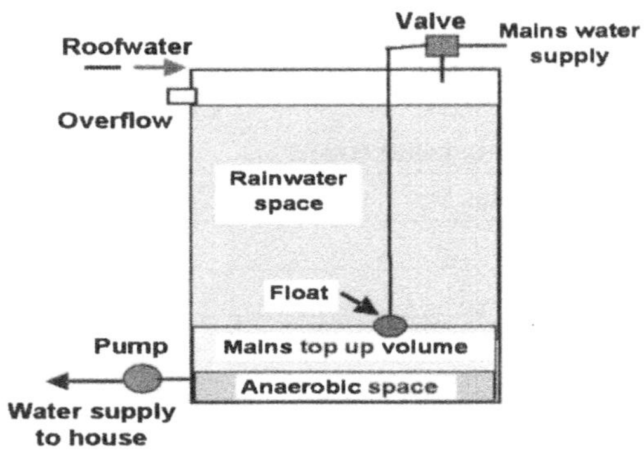

Figure 13.2.5: Rainwater tank details

The house is situated close to roads with high traffic loads and, unlike the Figtree Place and Maryville sites, trees containing birds overhang the house roof. A first flush device has not been installed and a small 125 L hot water service set at a maximum temperature of 60°C operates within the house. Water quality within the tank and at cold and hot water taps was found to the compliant with drinking water guidelines for metal, physical and elemental parameters. The average microbial results at different locations within the tank and at hot and cold taps are shown in Table 13.2.3.

Table 13.2.3: Average microbial water quality results at various locations

Location/ Criteria	Fecal Coliform (CFU/100 ml)	Total Coliform (CFU/100 ml)	Pseudomonas Spp. (CFU/100 ml)	Heterotrophic plate count (CFU/ml)
Water surface	108	1,050	3,100	1,050
Mid depth	34	900	780	427
Bottom	55	862	4,060	1,252
Cold tap	<1	200	412	76
Hot tap	0	2	<1	<1
Guideline	0	0	NA	200*

* *Japanese and American drinking water guidelines (Fujiwara et al., 1992)*

Table 13.2.3 shows substantial reductions in concentrations of bacteria in rainwater from the tank water surface to the hot and cold water taps. A natural rainwater treatment train is clearly evident. The quality of the hot water supply was compliant with drinking water guidelines with the exception of Total Coliforms. Interestingly, further analysis using Polymerase Chain Reaction (PCR) processes of rainwater samples that tested positive for Total Coliforms revealed that the water samples actually contained *Bacillus Spp.* (Spinks *et al.*, 2004). Further research by Evans *et al.* (2005; 2006) concludes that there is sparse evidence of faecal contamination in rainwater tanks. The lower temperatures, relatively nutrient poor and consequently highly competitive tank environments, would no doubt favour environmental organisms over enteric species adapted to the warmer, nutrient rich digestive tracts of animals. This result has significant implications for the assessment of the quality of rainwater using presence/absence of Coliforms in accordance with drinking water guidelines.

Significant reductions in concentrations of Pseudomonas Spp., heterotrophic plate counts, Fecal and Total Coliforms were observed between the bottom of the rainwater tank and the cold tap. It was assumed that high pressures in the pump also eliminated bacteria from rainwater supplies.

Water quality at the cold tap exceeded the guidelines for Fecal and Total Coliforms. The significance of exceeding drinking water guidelines for Coliform bacteria is unknown. Total Coliforms are no longer recognised as a suitable indicator of fecal contamination of water or having any relevance to health (Cunliffe, 2004). Although most studies find Coliform bacteria in rainwater storages and over 3 million Australians rely on rainwater for drinking water supplies only a small number of health concerns have been attributed to rainwater supplies. It is noted that the epidemiological study by Heyworth (2001) found that drinking rainwater posed a lesser health risk than drinking mains water in Adelaide Australia. A relationship between the presence of Coliform bacteria in rainwater tanks and frequency of illness could not be established.

Biofilmsin Rainwater Tanks

One of the dominant processes in the rainwater treatment train appears to be flocculation of organic, metallic and chemical parameters at the tank water surface with subsequent settlement of flocs to the sludge at the bottom of the tank or attachment to the walls of the tank. These flocs form biofilms at the interface between tank surfaces and stored rainwater potentially improving the quality of stored rainwater by removing contaminants.

A monitoring programme has been established to confirm the existence of biofilms in rainwater tanks and determine the impacts of biofilms on water quality in tanks. Preliminary monitoring by Piggott (2003) has targeted three different rainwater tanks as shown in Table 13.2.4.

One of the dominant processes in the rainwater treatment train appears to be flocculation of organic, metallic and chemical parameters at the tank water surface with subsequent settlement of flocs to the sludge at the bottom of the tank or attachment to the walls of the tank. These flocs form biofilms at the interface between tank surfaces and stored rainwater potentially improving the quality of stored rainwater by removing contaminants.

A monitoring programme has been established to confirm the existence of biofilms in rainwater tanks and determine the impacts of biofilms on water quality in tanks. Preliminary monitoring by Piggott (2003) has targeted three different rainwater tanks as shown in Table 13.2.4.

Table 13.2.4: Details of preliminary study into biofilms

Location	Material	Land use	Age (yrs)	Maintenance	Observed slime layer	Confirmed biofilm
Largs	Concrete	Rural	20	Frequent	None	Yes
Merewether	Plastic	Urban	3	Irregular	Slight	Yes
West Wallsend	Corrugated iron	Urban	14	None	Large	Yes

Table 4 shows that the location, construction material and level of maintenance relating to each tank is different. Nevertheless microbial testing of samples scraped from the walls of each tank has confirmed the existence of biofilms. Samples were taken from the walls of the tanks and PCR analysis used to identify the microbial organisms that make up biofilms in each tank as shown in Figure 13.2.6.

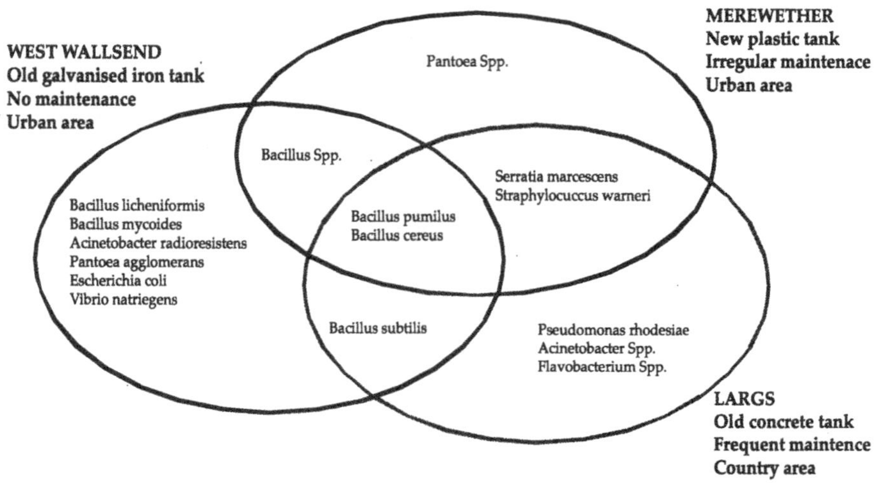

Figure 13.2.6: Distribution of microflora in biofilms in rainwater tanks at different locations

Figure 13.2.6 reveals that *Bacillus pumilus* and *Bacillus cereus* are common to the biofilms in all tanks whilst *Bacillus subtilis, Bacillus Spp.* (other than *Bacillus pumilus, Bacillus cereus* and *Bacillus subtilis*), *Serratia marcescens* and *Staphylococcus warneri* are common to the biofilms in two of the tanks. Biofilms in the rainwater tanks consisted of a core group of bacteria that are predominately *Bacillus Spp.* that originate from soils and the environment. These core bacteria isolated from the biofilms were not isolated in the stored rainwater in the tanks indicating that the process of forming biofilms may be removing bacteria from the tank water.

Elemental analysis using high resolution ICP-MS revealed accumulation of metals including lead, zinc, copper, manganese, chromium, mercury and arsenic in the biofilms in all tanks. Significant concentrations of metals were not found in the tank water indicating that the action of biofilms may be removing metals from the tank water.

Efficacy Of Bacterial Indicator Tests in Rainwater

The use of PCR methods to examine the DNA of bacteria sampled from rainwater harvesting systems revealed bacterial species that were not consistent with results implied by Coliform tests. Indeed the measurement of Fecal and Total Coliforms in rainwater samples indicated the potential presence of pathogens but subsequent analysis using PCR methods reveals that the bacteria was actually relatively harmless environment species such as *Bacillus Spp.* Given that the Coliform group of bacteria are used to indicate the safety of water supplies these potentially misleading results has profound implications for the utilisation of rainwater for domestic supplies.

Preliminary experiments were conducted to determine if bacteria other than E.Coli and Fecal Coliforms are isolated by commercial media providing misleading interpretation of rainwater quality. Two bacterial media that are approved by the USEPA were analysed namely: m-ColiBlue24® that is approved for isolating Total Coliforms and *E.Coli* from drinking water and m-FC broth with Rosolic Acid that is approved for isolating Fecal Coliforms from wastewater. Both media were inoculated with *E.Coli* from a human source, and *E.Coli, Bacillus cereus, Bacillus licheniformis, Aeromonas hydrophila* and Enterococcus *faecalis* from environmental sources. The resulting growth after incubation at 35°C for 24 hours is shown in Table 13.2.5.

The results from the preliminary experiment, shown in Table 13.2.5, confirmed that bacteria other than Fecal Coliforms, Total Coliforms and *E.Coli* can grow on commercially approved media. This indicates that the use of approved Coliform indicator tests can potentially result in the misleading view that rainwater supplies are unsafe. It should be noted that the results presented here are preliminary. More comprehensive analysis is needed and indeed necessary. Nevertheless this analysis does indicate the need for a comprehensive and critical review of the range of industry methods used to indicate the safety of rainwater supplies.

Table 13.2.5: Results of bacterial plate analysis

Bacteria	Present on Media	
	m-ColiBlue24	m-FC
E.Coli	Yes	Yes
Bacillus licheniformis	Yes	Yes
Bacillus cereus	No	No
Aeromonas hydrophila	Yes	No
Pseudomonas aeruginosa	No	No
Enterococcus faecalis	Yes	No

Elimination of Bacteria in Domestic Hot Water Services

The use of rainwater in hot water systems can have a significant impact on the viability of a dual water supply strategy that utilises rainwater (Coombes 2005; Coombes *et al.*, 2002). Continuing research by Spinks *et al.* (2004; 2003) has extended the finding that hot water services can eliminate bacteria by Coombes *et al.*, (2000; 2003) to analysis of the heat death behaviour of selected pathogens. The time to eliminate 90% of a bacterial population at a given water temperature (D-value) is shown for selected bacteria are shown in Table 13.2.6.

Table 13.2.6: Preliminary heat death results for selected bacteria (after Spinks et al. 2003)

Bacteria	D-value (seconds)		
	55°C	60°C	65°C
E. coli O3:H6 (wild type)	401	51	<2
E. coli O157:H7	223	67	3
Enterococcus feacalis (non-haemolytic)	901	131	19
Enterococcus feacalis (haemolytic)	633	77	7
Shigella sonnei biotype A	354	54	3
Pseuodomonas aeruginosa (wild type)	304	49	5
Serratia marcescens (wild type)	71	10	2
Salmonella typhimurium	77	4	<2
Aeromonas hydophilia	17	3	2
Klebsiella pneumoniae	22	<2	<2

Table 13.2.6 shows that potentially pathogenic bacteria (note that *E. coli* O3:H6 is not usually pathogenic) are eliminated rapidly from rainwater at temperatures of 60°C or greater. These results suggest that rainwater supplied to hot water services maintained at 60°C will provide water of adequate hygienic quality. These results extend the knowledge gained from analysis of the quality of rainwater in hot water services at the Figtree Place, Maryville and Carrington sites.

Conclusions

This paper has provided an overview of some key observations from the research program into the quality of water supply from rainwater tanks conducted over the last decade at the University of Newcastle in Australia. The research journey provides some key insights into water quality processes in rainwater tanks and highlights the need for continuing scientific endeavour to replace myths and agendas with facts about this important water source.

The importance of applied research into the performance of carefully monitored demonstration sites is established by the key observations from the Figtree Place, Maryville and Carrington housing projects. Monitoring of these projects revealed the existence of a rainwater treatment train that includes first flush devices, the rainwater tank and domestic hot water services. In addition establishment of these projects exposed many myths and assumptions about the quality of rainwater. Significantly, a sparsity of knowledge about the microbial processes in rainwater tanks was revealed.

One of the dominant processes in the rainwater treatment train appears to be flocculation of organic, metallic and chemical parameters at the tank water surface with subsequent settlement of flocs to the bottom of the tank or attachment to walls in the tank. Ongoing analysis, resulting from this initial observation, has revealed that biofilms do exist in rainwater tanks and that a core group of environmental bacteria such as *Bacillus Spp.* are likely to form biofilms in rainwater tanks.

Monitoring of the demonstration projects also led to a discovery that domestic hot water services set at temperatures greater than 52°C consistently eliminated bacteria from rainwater. This discovery led to laboratory experiments into the impact of hot water on the viability of selected pathogens. Potentially pathogenic bacteria were observed to be rapidly eliminated from rainwater at temperatures of 60°C or greater.

The use of Polymerase Chain Reaction (PCR) processes to determine the DnA of bacteria found in rainwater has increased concerns about the efficacy of the use of traditional coliform indicator organisms to determine the safety of rainwater supplies. Preliminary experiments confirmed that bacteria other than Fecal Coliforms, Total Coliforms and *E.Coli* can grow on commercially approved media. This indicates that the use of approved Coliform indicator tests can potentially result in a misleading view that rainwater supplies are unsafe.

Reference

Coombes P.J. (2005). Integrated Water Cycle Management: Analysis of resource security. *Water.* March edition. Australian Water Association. Sydney.

Coombes P.J, A. Spinks, C. Evans and R.H. Dunstan. (2004). Performance of rainwater tanks in the inner city during drought. *WSUD2004* Conference. Adelaide.

Coombes, P.J., Kuczera, G. and Kalma, J.D. (2003). Economic, water quantity and quality impacts from the use of a rainwater tank in the inner city, *Australian Journal of Water Resources,* 7(2), 101-110.

Coombes P.J., G. Kuczera, J.D. Kalma, and J.R. Argue (2002). An evaluation of the benefits of source control measures at the regional scale. *Urban Water.* 4(4) London, UK.

Coombes P. J., J.R. Argue and G. Kuczera (2000). Figtree Place: A case study in Water Sensitive Urban Development. *Urban Water.* 4(1). 335-343, London, UK.

Coombes P. J., G. Kuczera, J.D. Kalma and R.H. Dunstan, (2000). Rainwater quality from roofs, tanks and hot water systems at Figtree Place, 3rd *International Hydrology and Water Resource Symposium,* 1042-1047. Perth, Australia.

Coombes P. J., G. Kuczera, J.R. Argue, F. Cosgrove, D. Arthur, H.A. Bridgman and K. Enright, (1999). Design, monitoring and performance of the water sensitive urban development at Figtree Place in Newcastle. Proceedings of the 8th *International Conference on Urban Storm Drainage.* 1319-1326. Sydney. Australia.

Cunliffe, D. A. (2004). Guidance on use of rainwater tanks. EN Health Council. Department of Health and Ageing. Australian Government.

Evans C.A., P.J. Coombes and R.H. Dunstan, (2006). Wind, rain and bacteria: the influence of weather on the microbial composition of roof harvested rainwater. Water Research 40, 37-44.

Evans C.A., P.J. Coombes, R.H. Dunstan and T. Harrison, (2006). Identifying the Major Influences on the Microbial Composition of Roof Harvested Rainwater and the Implications for Water Quality. WSUD and UDM International Conference 1, 245-252. Melbourne. Australia.

Fujiwara M., Manwaring J.M., and Clarke R.M (1992). Drinking water in Japan and the United States: Conference Objectives. The proceedings of the 3rd U.S. – Japan Governmental Conference on Drinking Water Quality Management. Cincinnati, Ohio, USA.

Heyworth, J.E. (2001). A diary Study of Gastroenteritis and Tank Rainwater Consumption in Young Children in South Australia, 10th International Rainwater Catchment Systems Conference, Weikersheim, Germany, 1, 141-148.

Lloyd C., P. Troy and Schreiner S. (1992). For the public health. Longman Cheshire. Melbourne. Australia.

HMRC (1996) Australian Drinking Water Guidelines, Commonwealth of Australia, Canberra.

Pigott K. (2003). Biofilms in rainwater tanks. Honours Thesis. School of Environmental and Life Sciences. Department of Biological Sciences. University of Newcastle. Australia.

Prescott L.M., P.M. Harley and D.A. Klein (1999). Microbiology. 4th ed., McGraw Hill. USA.

Spinks A.T., R.H. Dunstan, T. Harrison, P.J. Coombes and Kuczera, G. (2004).Thermal inactivation of water-borne pathogenic and indicator bacteria at sub-boiling temperatures. Water Research 40, 1326 – 1332.

Spinks A.T., R.H. Dunstan, P.J. Coombes and G. Kuczera, (2003). Thermal Destruction Analyses of Water Related Bacteria in a Rainwater Medium at Domestic Hot Water System Temperatures. 28th *International Hydrology and Water Resources Symposium.* Wollongong. Australia.

Spinks A.T., P.J. Coombes, R.H. Dunstan and G. Kuczera, (2003a). Water Quality Treatment Processes in Domestic Rainwater Harvesting Systems. 28th *International Hydrology and Water Resources Symposium.* Wollongong. Australia

14 Aerobic and Anaerobic Treatment

14.1 Control of Mosquito in Domestic Septic Tanks by Larviciding

Nasiman Sapari[1], Muhammad Sanif Maulut[1] and S. Veeramohan[2]

University Technology Petronas, Malaysia[1]
Depatment of Health, Perak Tengah, Malaysia[2]

Abstract

Septic tank has been widely used for the treatment of domestic wastewater in urbanenvironment of Malaysia. This decentralize wastewater treatment system is a practical and effective way of treating the wastewater for houses in small residential areas. However, septic tanks in tropical environment posed a major problem due to the breeding of mosquitoes that include the aedes species, which are the vector of the dengue virus. This paper presents the problem and control of mosquitoes in septic tanks. Thirty septic tanks in three residential areas were examined in terms of design and problem of mosquito infestation. One hundred sixty septic tanks in two of the residential areas were treated with larvicide quarterly. In two weeks after the treatment, the results indicated that no mosquito or mosquito larvae in the treated septic tanks. However, the septic tanks that were not treated continuously become breeding ground for mosquitoes. Quarterly application of larvicide is an effective control of aedes mosquito in septic tanks under the tropical environment.

Key Words

Septic tank, larvicide, mosquito control, domestic wastewater.

Introduction

Mosquitoes are nuisance and hazard that can cause fatalities through the transmission of communicable diseases. Spraying of insecticide or fogging at targeted places commonly controls the adult mosquito population in urban environments. However a more effective way to maintain low mosquito population densities is by eliminating or controlling their breeding places. These places include construction sites, discarded containers in open spaces, and pockets of stagnant water in ground and engineering structures.

This paper examines the potential of septic tanks in becoming places for breeding of mosquitoes. Aedes mosquitoes, which can transmit dengue fever and dengue hemorrhagic fever, were found breeding inside the septic tank especially in the clear water retention section (Sivagnaname1 *et al.*, 2005)[1].

Control of dengue fever by fogging activities has not shown long-term success in controlling the cases (K B Chua, *et al.* 2003)2. In fact the number of dengue cases has been increasing every year in countries around the tropical and semi-tropical regions. Hence a more proactive approach is warranted instead of relying on reactive fogging.

Method of Study

This paper is the result of both field and laboratory studies. The field studies involved community participation, personnel from the Local Authorities and Federal Departments of Health in a joint effort in controlling dengue fever. Three residential areas were chosen in this study because of frequent reported cases of dengue fever. Two of the residential areas are in Sitiawan, Manjung Municipality while the other is in Batu Gajah Municipality. In Manjung, the houses chosen were single-storey link type and in Batu Gajah the houses were of the semi-detached type.

Septic tanks in the residential areas were examined in terms of their design and likelihood of retaining clear water that attracted mosquitoes to breed. The locations of the study areas are Sitiawan: 4°13'N, 100°42'E and Batu Gajah: 4°28'N, 101°03'E both having a warm and humid tropical climate.

Measurements of the septic tanks were carried out on various sections of the tank. Observations were also made on the following characteristics:

- The tank configuration
- The structural integrity
- Water tightness and openings
- The size of the tank and
- The various compartments inside the tank

Effluent samples were collected from the clear water section of the tank for quality analysis. The analysis was carried out in the PETRONAS University laboratory on the same day of sample collection. The parameters that were analyzed include BOD, COD, pH and suspended solids (SS). The type of adult and immature stages of mosquitoes and found in the tank were identified and recorded together with observations about the condition of the tank's cover and the level of water inside.

In their effort in controlling mosquitoes in septic tanks, the Manjung Local Authority in collaboration with the Department of Health, had directed the residents in their operational area to apply residual larvicides in their septic tanks. The procedure, as specified by the Department of Health involves the use of 250 g of sawdust tied into a cloth bundle. It is soaked for 24 hours in a 50 percent emulsion concentrate of Abate 500E™. The bundle of saw dust soaked in the larvicide is placed into the effluent retention section of the tank to ensure the slow release of larvicide which remains residual over a period of approximately three months.

Monitoring was carried out randomly in the 30 septic tanks two weeks after the application of residual larvicide. The septic tanks in the two residential areas in Sitiawan, Manjung Municipality were not fitted with pumps for the discharge of the effluent. As an alternative each tank has an outlet connected to a PVC pipe for discharge into the perimeter drain at the back of the house.

Results and Discussion

Examination of the configuration of the septic tanks in the three residential areas indicated that the septic tanks were constructed according to the conventional design. The filtration system had been expanded into a biofilter bed with large aggregate granite stones as the media. The effluent trickles uniformly over an inclined sheet of a perforated distribution plate onto the biofilm-coated aggregate and drains into a retention tank prior to discharge into the surface drain. Basically the septic tank is made up of three different compartments namely the septage, biofiltration and clear effluent tanks. The sizing of the tanks is shown in Table 14.1.1.

Table 14.1.1: The dimension and volume of septic tank's appurtenances

Appurtenance Dimension	(m×m×m)	Volume (m^3)
Septage tank	1.6×0.6×2.5	2.40
Biofiltration tank	1.4×2.5×2.5	8.75
Clear effluent tank	0.75×1.26×2.5	2.36

The plan and the cross-sectional views of the septic tank are shown in Figure 14.1.1. The tank is made up of reinforced concrete and each compartment is provided with a rectangular opening of 530 mm X 400 mm. The opening is covered with a cast iron plate of 3 mm thickness. The septage section was designed to retain the solids, which include the sludge and the scum. Trickling biofiltration process through a fixed media of about 2 m thick treats the clear liquor from the septage section. The effluent from the biofilter flows through the bottom of the biofilter into a retention tank for discharge using a submersible pumps activated by a ball and float switch.

The septic tanks in the two residential areas at Sitiawan are not fitted with such pumping facilities. The effluent is collected via gravity flow and accumulated discharge through a PVC pipe that connects the upper part of the tank to the surface drain. Results of the analysis of effluent samples from these septic tanks are shown in Table 14.1.2.

Table 14.1.2: Quality of the effluent from the septic tank

Parameter Concentration (mg/l except pH)	
pH	7.5
Chemical oxygen demand (COD)	57.3
Biological oxygen demand (BOD)	28.3
Suspended Solids (SS)	1.6
Turbidity	18.6

The above effluent quality could meet the discharge standard specified by the Department of Environment i.e., 50 mg/l for BOD and 100 mg/l for COD. The appearance of the effluent was generally clear.

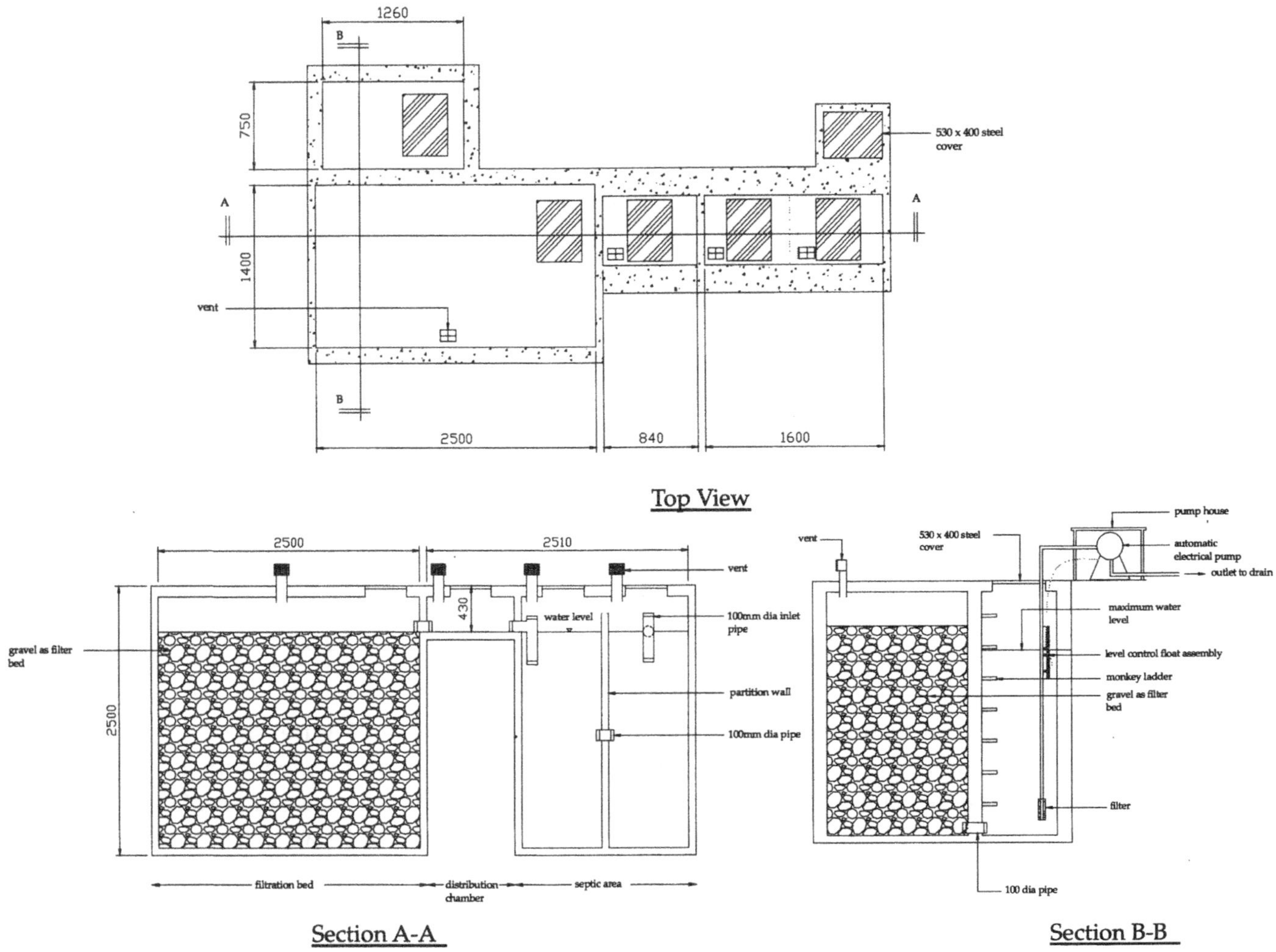

Note: All dimension are in mm units

Problem of Mosquito Infestation

The septic tanks in the study area are more than 10 years old and have problems related to pumping system malfunction. In the residential areas in Batu Gajah, only 10% of the populations maintain their septic tanks regularly. Some of the residents replace the older pumps with new submersible pumps. The malfunctioning of the pumps resulted in flooding of the filtration tank and water stagnation in the effluents tank. Investigation on the condition of the septic tanks revealed that all the flooded tanks were breeding Aedes *albopictus* mosquitoes (Figure 14.1.2). Mosquitos were observed entering the tanks mainly through the venting pipes and unsealed spaces between the manhole covers. It was observed in one septic tank that mosquitoes were escaping at almost regular intervals through a warped metal cover of a vent pipe.

A B

Figure 14.1.2: Infested septic tank - (A) Top view of the septic tank. (B) Photo of mosquitoes inside the septic tank.

It is a well-known fact that residential areas using septic tanks have higher mosquito populations (Mohd. Ridzwan and Nafizah3, 2006). However results of larvicide treatment showed a positive effect in preventing mosquito breeding in such tanks. All of the twenty septic tanks monitored in Sitiawan residential area showed no signs of mosquito infestation even though the water level in the tank was always high and in some cases the metal cover was completely missing.

Conclusion

The use of septic tanks is a convenient and cost effective system for the decentralized treatment of wastewater from individual houses in small residential areas. However, the system requires regular maintenance by the house owner. Otherwise it could lead to problems due to pump malfunction, which consequently causes the flooding of the filter, which can lead to mosquito breeding. Awareness of the system needs to be cultivated among house owners. Local authorities can play an important role in educating the public in this aspect. The application of larvicide for the control of mosquito breeding in septic tanks proved to be an effective method. Over a period of time, this approach can completely eliminate mosquito production in poorly maintained septic tank.

Reference

N. Sivagnaname, D. Dominic Amalraj & T. Mariappan. 2005. Utility of expanded polystyrene (EPS) beads in the control of vector-borne diseases, Indian J Med Res 122, pp 291-296.

K.B.,Chua I.L., Chua I.E and Chua, K.H. 2003. Effect of Chemical Fogging on Immature Aedes Mosquitoes in Natural Field Conditions. Singapore, Med. J. 46(11): pp 639-644.

Mohd. Ridzwan and Nasfiza M. 2006. Eradication of dengue: Engineering efforts. Paper presented at the 1st IFEH Asia and the Pacific Regional Conference on Environmental Health.

14.2 Settling Studies of SBRs Treating Dairy Effluent Operated at Industrial Scale

Joseph V THANIKAL[1,2], Michel TORRIJOS[1], Philippe SOUSBIE[1], Jean-Philippe DELGENES[1]

[1]INRA, LBE , Avenue des Etangs, 11100, Narbonne, France, torrijos@ensam.inra.fr

[2]Kumaraguru College of Technology, Coimbatore – 641 006, India

Abstract

From the mid of 90's, the SBR process has been applied in France to the treatment of cheese making plants, particularly in the Jura Mountains in north-east of France, where there are some 200 small cheese making plants. Each plant employs an average of 2 people and processes between 400,000 to 18,000,000 liters of milk a year, the average being 2,500,000 liters/year. Sludge samples were collected from these plants to study the performance of these installed SBRs. The sludge collected from 6 SBR plants was characterized for its settling properties. The results obtained were compared with the sludge from municipal wastewater treatment plants (WWTP), and showed that the SBR reactors treating dairy effluent have an excellent settleability and compressibility.

Key Words

Dairy effluent, SBR, settling, compressibility, WWTP.

Introduction

In the Jura and the Doubs *departments* (administrative entities) situated in the Jura Mountains in north-east France, there are some 200 small cheese making plants situated in sparsely populated rural areas. Subsequent to the setting up of a quality policy at the national level by the French cheese industry, the small production units in the Jura Mountains had to alter their installations in order to meet European health standards. Pollution from these cheese-making units is generally at the level of a 200-300 population-equivalent (P.E.) in hamlets with often as few as 100 to 150 inhabitants. For wastewater treatment, the cheese makers must choose between either an independent solution or being connected to the public sewage system for a collective treatment solution. Considering into account the large installation and operational cost of central sewerage system, sequencing batch reactors were suggested As a de-centralized treatment system for the dairy effluent. These reactors were designed for similar operating conditions and fabricated by the same fabricator. There are about 30 such SBR's are in operation in this region. Furthermore, it has been shown that the sludge that develops in SBRs treating dairy effluent at laboratory scale or industrial scale have excellent settling characteristics (torrijos 2001 and torrijos 2004). This paper deals with the characterization of sludge from 6 full-scale SBRs from the Jura mountains, which represents 20 % of the existing SBRs of this area. The aim of the work was to measure the settling characteristics of the sludge of these SBRs to check: (i) first, if the sludge which develops in the SBRs keep, after several years of operation by the dairy staff, characteristics close to the characteristics obtained in shorter experiments at laboratory or full-scale; (ii) second to check if the sludge allow a good operation of the SBRs.

Materials and methods

Sequencing batch reactor process

The wastewater, given its concentration of organic matter, is treated in a low loaded aerobic reactor working as an SBR, at one treatment cycle a day. The effluent generated during one working day is held temporarily in a first buffer tank. At the start of the cycle, the reactor containing the activated sludge is filled with the effluent to be purified (fig. 14.2.1) and then must be continuously aerated for 20 hours, followed by decanting, which lasts 3 hours. Then, the volume initially introduced is run off and, when necessary, a portion of the sludge removed from the bottom of the reactor tank. 4 hours after the end of the aeration stage, a new cycle can be started up.

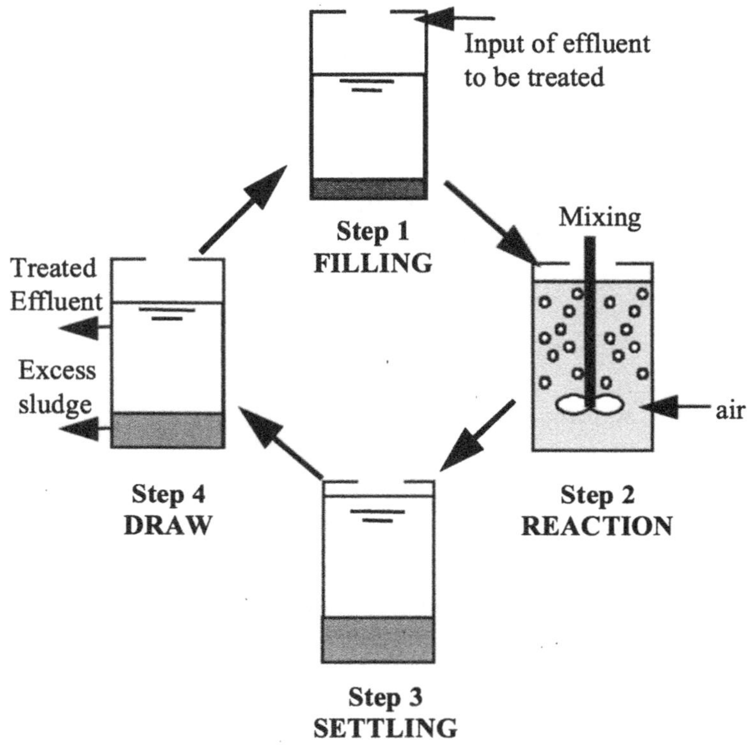

Figure 14.2.1: The operating principle of an SBR process for the treatment of cheesemaking effluent.

Sludge samples collection

The sludge samples were collected from the SBR tank at the end of the aeration phase, just before settling started. Transportation time between the SBR plants and the laboratory was 7 hours. Sludge were maintained at ambient temperature during transportation and refrigerated to 4°C as soon as they reached the laboratory. Test were completed within 48hrs. The 6 samples collected were identified as : A, Ar, F, G, N and O according to the first letter stands for the village where they were collected.

Sludge volume index and zone settling velocity (SVI & ZSV)

The test was carried out in a transparent cylinder of 1 L and 10 cm in diameter as per standard methods, APHA (1992). The sludge was diluted, when the settled volume after 30 minutes exceeded more than 25% volume of the cylinder.

The settled sludge volume was measured at every 5 minutes up to 30 minutes. The SVI was expressed as settled sludge volume in ml/g of MLSS. The slope of the initial straight line portion of the curve plotted between time for settling and settled sludge volume gives the ZSV. To simulate the behavior of the sludge in the SBR, a settling test was also carried out with non-diluted sludge and the height of the settled sludge was measured after 30 minutes, 1, 2 and 4 h.

Flocculating ability

Flocculating ability of the sludge was measured as the re-flocculation ability of the disrupted sludge by the method described by Jorand *et al.* (1994). The sludge was disrupted by ultrasound using the sonication probe provided with Beckman Coulter LS200 granulometer. Initial tests were carried out, to fix the conditions of sonication with different sonication power and time of sonication. 80 ml of sludge sample, with a MLSS concentration of 4 g/L, were sonicated at 70 watts for 30 s. 10 ml aliquot was centrifuged at 125 x g for 2 minutes after the Sonication and the absorbance of supernatant, designated as A, was measured at 650 nm, using an UV spectrophotometer. The remaining sample was agitated using a paddled impeller at a minimum speed ,such that the solids were just kept in suspension. After 15 minutes of agitation, 10 ml aliquot was analyzed for absorbance as before and designated as B. Tests were conducted in triplicate. Flocculating ability was calculated as follows:

$$\text{Flocculating ability (\%)} = (1 - B/A) \times 100$$

Floc size distribution

Floc size distribution was measured by using a Beckman Coulter LS200 granulometer. The instrument uses Fraunhofer method of diffraction of light for measurement of size of floc in the range of 0.2 – 2000 μm. The samples are added to vessel containing water, which is provided with sensors, detecting the % concentration of sample. The addition of sample is done until the concentration is between 9-11%. The light scattered is detected by detector that converts the signal to a size distribution based on volume.

Results

Characteristics of the raw wastewater

The effluent to be treated is the water used for cleaning the tanks, the pipes, the cheese moulds and the wash water from the floor of the cheese dairy. The sweet whey obtained during the forming, transfer, and palletisation during the first pressing stages and the acidic whey obtained at the end of pressing, are carefully recovered and kept apart from the rinsing water.

Table 14.2.1: 1 : COD at inlet and outlet of the 6 SBRs.

	Reference of the SBR plants					
	A	Ar	F	G	N	O
Raw wastewater COD [(1)]	4.3 g/L (+/- 1.78)	5 g/L (+/- 1.35)	1.6 g/L (+/- 0.54)	4.2 g/L (+/- 2.05)	1.8 g/L (+/- 1.53)	1.5 g/L (+/- 0.72)
Treated wastewater COD [(2)]	48 mg/L	133 mg/L	18 mg/L	312 mg/L	70 mg/L	26 mg/L

[(1)] *Results from the independent control office.*

[(2)] *Results from the isolated outlet sampling carried out the day of sludge sampling.*

Table 14.2.1: 1 : COD at inlet and outlet of the 6 SBRs.

Parameter	Unit	Reference of the SBR plants					
		A	Ar	F	G	N	O
SS concentration	g/L	2.06	6.18	4.8	5.72	10.35	2.85
VSS/SS	%	71	85	72	68	66	61
SVI	mL/g	63	153	54	140	72	42
ZSV	m/h	3.6	3.1	3.9	1.2	3.7	4
Height of sludge after 4 h of settling	mL/L	110	260	220	510	645	110
Non diluted SVI	mL/g	53	42	46	89	62	39
Floc size	µm	210	86	65 - 300	54	107	175
Flocculating ability	%	33	56	40	35	56	50

Sludge settleability (SVI and ZSV)

Four samples A, F, N, O had a low value of SVI, between 50 and 75 mL/g. These values were in accordance with the values of SVI normally observed for SBR sludge treating dairy effluent (Torrijos,2001). However, the samples G and Ar, showed higher SVI values with, 140 and 153 ml/g respectively. These values were close to the SVI measured with activated sludge treating municipal wastewater (Brite-Marie, 2003).Extensive pilot plant studies, treating municipal wastewater by G.T. Daigger, R.E. Roper, Jr., showed that the SVI values ranged from 36 – 402 mL/g.

The sludge G was the only one with a very low ZSV (1.19 m/h) when the 5 other sludge had quite close ZSV with values between 3 and 4 m/h which correspond to the highest values found in domestic wastewater treatment (britt marie , G.T. Daigger, R.E. Roper, Jr.).

In addition of the standard measurement of the SVI after 1/2 hour, the height of the un-diluted sludge in a 1L cylinder after 4 h of settling was also measured to fit to the real conditions of operation of the SBRs. Indeed, to operate the SBR and, above all, to manage the selective withdrawal of sludge from the reactor, the cheese makers estimate, every week, the height of the sludge in the reactor after settling, by measuring the volume of the un-diluted sludge in the 1 L cylinder after 4 hours. When the level exceeds a predetermined volume, the plant operator has to withdraw the sludge at the end of settling to decrease the sludge concentration in the reactor.

The comparison of the height of the sludge in the 1L cylinder after 4 hours of settling and the sludge concentration leads to the following conclusions: (i) SBR G is the only one sample where the sludge behavior is bad with a settling of half the volume of the reactor with a suspended solids concentration 5.7 g/L. Hence, the sludge volume index is quite higher than for the five other sludge (40 to 60 mL/g) ; (ii) Sludge should be removed from SBR N because the level of sludge is high due to a high SS concentration (10.35 g/L). However the settling characteristics of this sludge were quite good with an un-diluted SVI close to 60 mL/g.

Flocculating ability

The investigations showed that the level of sonication required to disrupt the flocs from SBR treating diary effluent was different (70 W, 30 s) from that used for effluent from WWTP's, (50 W, 15 s., Britte Marie 2003). In the present study the flocculating ability for sludge samples varied from 30 – 57%. The sludge samples A, F and G showed a low flocculating ability (30 – 40%) when compared with samples Ar, N and O (50 – 57%). This may be explained as, samples of A and F contained Flocs with larger size. A microscopical examination revealed that these flocs were formed with filamentaceous bacteria. Flocs could not re-flocculate easily due to fragmentation. Sludge G had very disrupted flocs which can explain a very low ability to flocculation. The floc size was comparatively smaller in other samples (Ar, N, O), which could re-flocculate easily. Britte- Marie (2003) made similar observation) with municipal wastewater. There was no significant correlation between flocculating ability and SVI. However irrespective of the flocculating ability all the samples showed good settling properties (SVI <150 ml/g).

Floc size distribution.

The floc size determined using a Beckman Coulter apparatus showed different particle size distribution for the various sludge. Out of the six samples analyzed, samples A and F showed a wide particle size distribution varying from 0.4 to 2000 μm. The other samples (Ar, G, N and O) were present with floc sizes varying from 0.4 to 600 μm, showing a more homogeneous distribution. The particle size distribution for municipal wastewater had generally values less than 600 μm, (Britt-Marie, Ruxandra (2003)). The present study showed that the sludge with larger flocs had good settling properties. However floc size varied in various sludge samples. Indeed, The results showed that the samples (A, F, N, O) with floc size > 100μm had a SVI < 100mL/g showing good settling property. Results of municipal wastewater (Britt-Marie 2003) showed that the samples with smaller floc size had good settling and the samples with larger diameter of flocs did not settle well. This may be due to the fact that, the flocs with larger diameter were not dense but shown a larger surface area. The same criteria was also observed by Andeadakis (1993).

Conclusion

The sludge from 6 SBRs, operated at industrial scale, were collected to study the settling of sludge in the reactors. The studies lead to the following conclusions. All the samples showed a satisfactory undiluted SVI after 4 hour of settling indicating that the SBR could be operated correctly in these conditions with a good liquid-flocs separation. A direct correlation between flocculating ability and floc size could not be established, but it was found that the samples with larger floc size showed low flocculating ability. Samples with larger floc size shown good SVI. However the samples with smaller floc size showed better re-flocculating ability after disruption. This may be due to, the smaller flocs were stable and was not fragmented further.

The study showed that the sludge settling was good in the SBRs. Care must be taken to recover sweet whey before letting the effluent into the SBR, in order to avoid poor performance linked to overloading of the reactor. The studies also conclude that 4 sludge had characteristics close to the characteristics obtained at lab or full-scale experiments but that 1 had quite different characteristics linked to a bad operation of the reactor by the cheese maker and repetitive overloadings of the reactor.

References

Andeadakis A., 1993. Physical and Chemical properties of activated sludge floc, *Wat. Sci. Tech..* 27 (2), 1707-1714.

APHA, Standard methods for the examination of water and wastewater, 19th edition, American public health association Baltimore, MD, 1995.

Daigger G., Roper R., 1985. The relationship between SVI and activated sludge settling characteristics, *Journal WPCF*, vol 57, 859-866.

Govoreanu R., 2004. Activated sludge flocculation dynamics: On-line measurement methodology and modeling, PhD thesis, universiteit Get, Belgium.

Jenkins D., Richard M., Daigger G., Manual on the causes and control of activated sludge bulking and foaming, Ridgeline press, Lafayette, CA 1993.

Jin B., Wilen B-M. and Lant P., 2003. A comprehensive insight into floc characteristics and their impact on compressibility and settelability of activated sludge, *chemical engineering journal*, vol 95, 221-234.

Sezgin M., Variation of sludge volume index with activated sludge characteristics, Water Res. Volume 16 (1982) 83-88.

Torrijos M., Vuitton V., Moletta R., 2001. The SBR process: an efficient and economic solution for the treatment of wastewater at small cheese making dairies in the Jura mountains. *Wat. Sci. Tech.*, 43 (3), 373-380.

Torrijos M., Sousbie P., Moletta R., Delgenes J-P., 2004. High COD wastewater treatment in an aerobic SBR : treatment of effluent from a small farm goat's cheese dairy. *Wat. Sci. Tech.*, 50 (10), 259-267.

IWA Publishing's authorised EU representative for General Product Safety Regulations is Diane D'Arras, 15 rue Duret, 75116 Paris, France, e-mail: safety@iwap.co.uk.

Printed and bound by CPI Group (UK) Ltd, Croydon, CR0 4YY
12/05/2026
02108892-0016